T0135605

Rheology and Dynamics of Simple and Complex Liquids in Mesoporous Matrices

Dissertation
zur Erlangung des Grades
des Doktors der Naturwissenschaften
der Naturwissenschaftlich-Technischen Fakultät II

– Physik und Mechatronik –

der Universität des Saarlandes

von

SIMON ALEXANDER GRÜNER

Saarbrücken

2010

Bibliografische Information der Deutschen Nationalbibliothek

Die Deutsche Nationalbibliothek verzeichnet diese Publikation in der
Deutschen Nationalbibliografie; detaillierte bibliografische Daten sind
im Internet über http://dnb.d-nb.de abrufbar.

ISBN 978-3-8325-2596-5

Logos Verlag Berlin GmbH
Comeniushof, Gubener Str. 47,
10243 Berlin
Tel.: +49 (0)30 42 85 10 90
Fax: +49 (0)30 42 85 10 92
INTERNET: http://www.logos-verlag.de

Tag des Kolloquiums: 21. Juli 2010

Dekan: Univ.-Prof. Dr. rer. nat. C. Becher

Mitglieder des
Prüfungsausschusses: Univ.-Prof. Dr. rer. nat. H. Rieger (Vorsitz)

Priv.-Doz. Dr. rer. nat. P. Huber (Gutachter)

Univ.-Prof. Dr. rer. nat. R. Pelster (Gutachter)

Univ.-Prof. Dr. rer. nat. O. Paris (Gutachter)

Dr. rer. nat. H. Wolf (Akademischer Mitarbeiter)

Affidavit

I hereby swear in lieu of an oath that I have independently prepared this thesis and without using other aids than those stated. The data and concepts taken over from other sources or taken over indirectly are indicated citing the source. The thesis was not submitted so far either in Germany or in another country in the same or a similar form in a procedure for obtaining an academic title.

Saarbrücken, March 9th, 2010

Simon Alexander Grüner

Abstract

Subject of my thesis is a study of rheologic and dynamic properties of fluids confined in an isotropic pore network with pore radii of $\sim 5\,\mathrm{nm}$ embedded in a monolithic silica matrix (porous Vycor®). The experimental technique bases on the capillary rise of a wetting liquid in a porous substrate, also known as spontaneous imbibition.

A crucial part of the conducted experiments centers on the increasing relevance of the liquid-substrate interface in the mesopore confinement. Detailed analyses of the measurements carried out with water, silicon oils, and a series of hydrocarbons result in precise information on the boundary conditions expressed in terms of the velocity slip length. Systematic variations of the chain-length of the used hydrocarbons also allow for an assessment of the influence of the shape of the liquid's building blocks on the nanoscopic flow behavior. Supplemental forced throughput experiments additionally address the impact of the liquid-substrate interaction.

Furthermore, the influences of spatial confinement on the surface freezing transition of the linear hydrocarbon n-tetracosane as well as on the mesophase transitions of the liquid crystal 8OCB are investigated. Finally, a third, more general study focuses on the kinetics of the invasion front, which is supposed to be influenced significantly by the random environment of the pore space considered.

Kurzdarstellung

Das Themengebiet meiner Dissertation umfasst rheologische und dynamische Eigenschaften von Flüssigkeiten in einem isotropen Porennetzwerk einer monolithischen Glasmatrix (poröses Vycor®). Die Porenradien betragen dabei etwa $5\,\mathrm{nm}$. Die durchgeführten Experimente beruhen auf dem Prinzip des Kapillarsteigens einer benetzenden Flüssigkeit in einem porösen Substrat (spontane Imbibition).

Ein Schwerpunkt der Untersuchungen liegt auf dem zunehmenden Einfluss der Grenzfläche zwischen Fluid und Substrat als Folge der extremen, räumlichen Beschränkung. Analysen der Experimente mit Wasser, Silikonölen und einer Reihe Kohlenwasserstoffe liefern präzise Angaben zur hydrodynamischen Randbedingung beschrieben durch die sog. Schlupflänge. Systematische Variationen der Kettenlänge der verwendeten Kohlenwasserstoffe erlauben die Beurteilung der Bedeutsamkeit der Molekülform für das mikroskopische Fließverhalten. Dank einer alternativen Experimentführung kann man sich auch mit dem Einfluss der Benetzbarkeit der Flüssigkeit befassen.

Darüber hinaus wird der Einfluss der räumlichen Beschränkung auf Phasenübergänge behandelt. Im Speziellen werden das Oberflächengefrieren eines Alkans und die Mesophasen eines Flüssigkristalls untersucht. Schließlich wird die Aufrauung der voranschreitenden Benetzungsfront beim Kapillarsteigen in dem mesoporösen Glas untersucht. Man erwartet, dass deren Bewegung durch die zufällige Netzwerk-Topologie entscheidend beeinflusst wird.

Was ist das Schwerste von allem?
Was Dir am Leichtesten dünkt:
Mit den Augen zu sehen,
was vor den Augen dir liegt.
(Johann Wolfgang von Goethe)

Contents

Symbols

Here the reader might find frequently used symbols along with their meaning and page references for more detailed information.

symbol	meaning [page]
A	sample cross-sectional area [5]
Bo	Bond number [21]
b	slip length [8]
β	growth exponent [59]
C	capacitance [96]
C_{cal}	MFA calibration factor [97]
Ca	capillary number [28]
c	washout parameter [44, 64]
D	diffusion constant [123]
d	sample thickness [7]
Δp	pressure difference [5]
η	dynamic viscosity [5]
η_1, η_2, η_3	Miesowicz viscosities [84]
f	filling fraction [68, 120]
Γ	imbibition strength [24] (Γ_{c} [64], Γ_{lb} [62], Γ_{m} [40], Γ_{r} [44], Γ_{t} [41], Γ_{ub} [62])
γ	interfacial tension [13]
γ_{C}	critical surface tension [13]
h_0	sample height [41]
$h(t)$	rise level [24]
$h_f(t)$	rise level of a certain filling degree f [69]
I_∞, I_0	profile levels [45]
K	membrane permeability [7, 10]
Kn	Knudsen number [124]

$\mathcal{L}$ characteristic length scale [7, 21, 124]
ℓ all-trans molecule length [50]
λ mean free path [124]
λ_c capillary length [19]

M sample's overall mass uptake [22]
$m(t)$ mass increase [24]

P reduced vapor pressure [120]
$P(r)$ pore size distribution [11]
p_L Laplace pressure [19]
ϕ_0 volume porosity [5]
ϕ_i initial volume porosity [21]

Re   Reynolds number [7]
r pore radius [5, 70]
r_0 mean pore radius [5, 12, 24]
r_h hydrodynamic pore radius [8, 24]
r_L Laplace radius [22, 24]
ρ volume density [7]

$\mathcal{S}$ gas solubility in a liquid [103]
σ surface tension [19]
σ_{tot} total scattering cross section [35]

T_c clearing point [83]
T_f bulk freezing temperature [75]
T_m pore freezing temperature [80]
T_s surface freezing transition temperature [75]
t_0 filling time [41]
θ_0 static contact angle [13]
θ_D dynamic contact angle [27]
τ tortuosity [6]
τ_r relaxation time [123]

$\dot{V}$ volume flow rate [5]
V_s sample volume [22]
v advancement speed [8, 28]
v^n normalized imbibition speed [78]

$w(t)$ vertical front width [59]

x_0 inflection point parameter [44]

1. PREFACE

"There's Plenty of Room at the Bottom". Ever since Richard Feynman's visionary talk on nanotechnology in 1959 great advances in the field of observation and manipulation of matter down to the atomic scale have been achieved. A variety of newly invented fabrication techniques comprising both top-down and bottom-up approaches have thereby been arising. Nowadays *micro-* and *nano-* applications are part and parcel of our everyday life, ranging from the titanium dioxide particles in sun milk to the accelerometer in airbags, mobiles, and digital cameras.

Apart from this, the research area of micro- and nanofluidics gains more and more importance. The development and the design of miniaturized structures and complex micro-machines, through which fluids move, are already forged far ahead. These days, manipulating and analyzing tiny amounts of a liquid in a lab-on-a-chip device is a biomedical standard implementation. Chemical reactions taking place in microreactors are the subject of worldwide academic research. And, last but not least, ink-jet printers wield ink drops of a few picoliters only.

The tendency of miniaturization is comparable to the ongoing developments in semiconductor industries. But pushing toward the world of atoms is accompanied with the occurrence of seemingly new physical phenomenons. Geometrical restrictions on the micro- and nanometer scale entail a variety of astonishing confinement effects. Furthermore, miniaturized setups go hand in hand with a gradual increase of the surface to volume ratio. Hence, interfacial effects, which have so far been negligible on the macroscale, may now affect or even dominate the fluid behavior.

Therefore, it is evident that the ongoing diminishment of the flow geometries requires profound knowledge of the impact of confinement and interfacial effects. In this regard, especially examinations of thin liquid films and a series of molecular dynamics simulations have been performed. However, to date there are only few experimental studies on the influence of confinement on the fluid dynamics.

In this context my thesis comprises a systematic analysis of the restriction-influenced flow behavior of a wide variety of different liquids in mesoporous matrices. By definition such substrates contain geometrical restrictions on the order of some 10 nm. In such geometries the fluid dynamics are characterized by both confinement and interfacial effects. Thus, they can also be simultaneously studied. In this regard observations of mesoscopic flow dynamics allow one to bridge the gap between continuum flow concepts and the molecular motion of single molecules.

Outline

Part I supplies a theoretical framework for most of the investigations presented subsequently. The focus is laid on mathematical conceptions aiming at a description of flow through a complex pore network of mesopores. The applied porous matrices are introduced as well.

Part II is the most extensive part of this thesis. I will present a systematic study of the self-propelled invasion dynamics of liquids in mesoporous matrices, also known as spontaneous imbibition. Prior to introducing the applied measuring methods the underlying physical mechanisms are discussed in detail. The subsequent experimental part considers the invasion dynamics of a series of liquids, in particular with regard to the interfacial behavior and the shape of the liquid's building blocks. What is more, the influence of confinement on the phase transition behavior is studied for one hydrocarbon and one liquid crystal. A third, more general study focuses on the kinetics of the invasion front, which is supposed to be significantly influenced by the random environment of the pore space considered.

Part III presents forced imbibition measurements that are, in principle, complementary to the study of the rise dynamics in Part II. Here, in contrast to the spontaneous (self-propelled) imbibition, the liquid flow is driven by an externally applied pressure gradient across the mesopores, though. This method particularly renders possible examination of non-wetting fluids.

In Part IV the results of the presented investigations are summarized. Moreover, some conceivable future studies, initiated by results gained within this thesis, are considered.

Finally, the Appendix comprises a data collection of relevant properties of the investigated liquids and gives a general overview of the methods applied for the matrix characterization.

Part I.

Introduction

Part I of this thesis will give some basic information on the description of the liquid flow through a complex network of pores on the mesoscale. Special attention will be paid to concepts regarding the consideration of the detailed network structure. On top of this, the influences of the mesopore confinement and consequent doubts about the applicability of standard assumptions of classical hydrodynamics will be addressed. Finally, the applied mesoporous matrices and their properties will be thematized.

2. Principles of Liquid Flows

Most measurements presented within this thesis involve the flow of a liquid through a complex pore network comparable to a sponge, namely porous Vycor® glass. In this porous glass the mean pore radii are on the order of a few nanometers only. At first glance, this seems to pose a rather sophisticated problem. Nonetheless, as a simple approximation one can reduce the problem to its fundamental phenomenon that is the flow of a fluid through a tiny capillary. Consequently, the law of Hagen-Poiseuille should give a not too bad description of the basic phenomenology. It is the starting point of the subsequent development of a theory of the liquid flow in a pore network.

2.1. Liquid Dynamics in Isotropic Networks

The flow of a liquid through a pipe is a prime example for the direct application of the Navier-Stokes equation resulting in the famous law of Hagen-Poiseuille. For a given pressure difference Δp applied along a cylindrical duct with radius r and length ℓ the volume flow rate $\dot{V}$ is determined by

$$\dot{V} = \frac{\pi r^4}{8 \eta \ell} \Delta p \, . \tag{2.1}$$

Here η denotes the dynamic viscosity of the flowing liquid. In the next step one has to evolve concepts in order to account for the sponge-like structure of an isotropic pore network. For this purpose a set of quantities is required that permits a sufficient characterization of such a matrix. The most prominent approach is the concept of the tortuosity τ, which will be introduced in the following.

2.1.1. Characterization of a Pore Network

In general an isotropic pore network can be characterized by three quantities. The mean pore radius r_0 and the volume porosity ϕ_0, as deduced from sorption isotherm experiments (see Appendix B), are probably the most intuitive ones among them. With only these two parameters a porous cuboid with edge length a (and cross-sectional area $A = a^2$) consisting of

$$n = \frac{\phi_0 A}{\pi r_0^2} \qquad \left(\Leftrightarrow \phi_0 \equiv \frac{V_{\text{void}}}{V_{\text{sample}}} = \frac{n \pi r_0^2 a}{a^3} \right) \tag{2.2}$$

cylindrical pores with radius r_0 and length a can be constructed. Assuming the capillaries to be aligned in flow direction the flow rate through the whole matrix is then given by n times the single pore flow rate Eq. (2.1) with $r = r_0$ and $\ell = a$.

However, so far this description still lacks information on the orientation of the pores.

To account for the isotropy of the network as indicated in Fig. 2.1 (left) it is necessary to introduce a third parameter, the so-called tortuosity τ along with the transformation

$$\dot{V} \;\longrightarrow\; \frac{1}{\tau}\dot{V} \tag{2.3}$$

of the volume flow rate Eq. (2.1). Pores totally aligned in flow direction would yield $\tau = 1$, whereas isotropic distributed pores would result in $\tau = 3$. This can be seen very easily. For a random orientation only every third pore is subjected to the pressure gradient and hence contributes to the flow. Therefore the net flow rate has to be divided by the factor three. But no correction is needed if all pores are aligned in flow direction and as a result of this it is $\tau = 1$. In this way the tortuosity is a simple method for accounting for the orientation of the pores with respect to the direction of the pressure drop.

To date serveral techniques have been applied to extract the tortuosity of the isotropic pore network in Vycor® glass. Deducing the diffusion coefficient of hexane and decane by means of small angle neutron scattering (SANS) measurements τ was found to be in the range of 3.4 - 4.2 [1]. Gas permeation measurements performed with an in-house apparatus resulted in $\tau = 3.9 \pm 0.4$ [2]. Finally, calculations based on three-dimensional geometrical models yielded a value of approximately 3.5 [3].

Interestingly, all values show a significant deviation from $\tau = 3$ as derived from the previous considerations. Accordingly, there must be an additional aspect of the geometry that has so far been neglected. Regarding Fig. 2.1 (right) this issue is apparent: the pores are not straight but rather meandering. In consequence

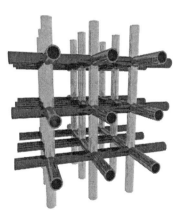 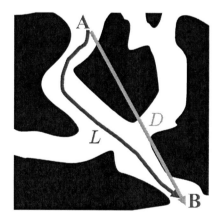

Figure 2.1.: Illustration of the meaning of the tortuosity τ of a pore network such as Vycor®. (left): For an isotropic distribution of the pores only every third pore is subjected to the pressure gradient yielding $\tau = 3$. (right): For meandering pores an additional factor of $\frac{L}{D}$ must be introduced to correct the length D of the direct interconnection of two points for the actual path length L.

the length L of the path from any point A to another point B is always larger than the length D of the direct interconnection of the two points. To correct the pore length for the larger flow path an additional factor of $\frac{L}{D}$ for the tortuosity must be introduced. Assuming $\tau = 3.6$ this consideration yields for the Vycor® pore network $L \approx 1.2\,D$. This result can vividly be interpreted as follows: the shortest way from the bottom of the previously introduced sample cuboid to its top is about 20 % longer than its edge length a.

2.1.2. Darcy's Law

With all the preceding considerations in mind one is able to conclude an expression that describes the flow of a liquid through a porous network. For a given porous matrix with cross-sectional area A and thickness d (along which the pressure drop Δp is applied) the normalized volume flow rate $\frac{1}{A}\dot{V}$ is determined by

$$\frac{1}{A}\dot{V} = \frac{K}{\eta d}\Delta p \,. \tag{2.4}$$

This expression is also known as Darcy's law [4]. The proportionality constant K is the so-called hydraulic permeability of the matrix. It is given by

$$K = \frac{\phi_0}{8\tau}r_0^2 \,. \tag{2.5}$$

At this point it must be emphasized that the permeability is solely specified by the matrix' internal structure and consequently it should be independent of the liquid and of the temperature.

2.2. Influence of the Confinement

So far I have completely neglected that the mean pore diameters of the pore network are orders of magnitude smaller than characteristic in usual flow paths in common miniaturized fluid manipulating applications. Indeed, the pore radii are merely 10 to 100 times larger than typical molecular diameters of simple liquids like water. Within the systematic study of chain-like hydrocarbons, which will be presented in section 6.3, the dimensions of the liquid's building blocks even exceed the channel's diameter. For that reason it is evident that some questions about the influence of the confinement on the fluid dynamics arise. In the following the two most apparent ones will be discussed.

Though, beforehand, I will point out a remarkable feature being inherent in micro- and nanofluidic devices and applications. Because of the tiny characteristic length scales $\mathcal{L}$ the Reynolds number

$$\mathrm{Re} \equiv \frac{\text{inertial force}}{\text{viscous force}} = \frac{\rho v \mathcal{L}}{\eta} \tag{2.6}$$

(with the liquid's density ρ) usually fulfills $\mathrm{Re} \ll 1$. Thus, flows in such restricted geometries are most likely laminar rather than turbulent. From this point of view the implementation of the Hagen-Poiseuille law seems to be justified. But, for very high flow speeds v this assumption does not remain valid. In section 4.6 it will be shown that this fact plays a crucial role in the very initial phase of a capillary rise experiment.

2.2.1. Validity of Continuum Mechanical Theory

Up to now I have assumed the law of Hagen-Poiseuille to be valid even in pores with diameters below 10 nm. However, one must not forget that this law is based on the principles of continuum mechanical theory, in which the behavior of a fluid is determined by collective properties such as the viscosity η and the surface tension σ. This assumption certainly holds for ensembles of 10^{23} molecules. But within the pore confinement such amounts are not reached. This can easily be seen in the following example. Assuming water molecules to be spheres with a radius of 1.5 Å in a hexagonal close-packed structure one arrives at only 1000 molecules per cross-sectional area. As a consequence, the validity of the continuum theory has to be put into question.

On this score especially the development of the surface force apparatus (SFA) has stimulated extensive studies over the last three decades. The mobility of water and several hydrocarbons in extremely confined films was examined by experiment [5–7] and in theory [8]. These studies revealed a remarkable robustness of the liquids' fluidity down to nanometer and even subnanometer spatial confinement. Moreover the validity of macroscopic capillarity conceptions at the mesoscale was demonstrated [9, 10]. The measurements within this thesis will provide further hints whether the concepts of viscosity and surface tension still remain valid in mesopore confinement.

2.2.2. Validity of the No-Slip Boundary Condition

The law of Hagen-Poiseuille implies the no-slip boundary condition. This means that the velocity of the fluid layers directly adjacent to the restricting walls equal the velocity of the walls themselves. Nowadays it is indisputable that this assumption does not hold unreservedly. Already 60 years ago Peter Debye and Robert Cleland introduced both slipping and sticking fluid layers at the pore walls in order to interpret a seminal experiment on liquid flow across porous Vycor® [4]. In that way, they were able to quantitatively account both for increased as well as for decreased measured flow rates (compared to the predicted ones) within their examinations of the flow of hydrocarbons through porous Vycor®.

The concepts of a sticking and of a slipping liquid compared to the traditional no-slip boundary condition are exemplified in Fig. 2.2 for a cylindrical tube with radius r_0. The degree of slip can be quantified by the slip length b with $r_0 \equiv r_h - b$. The hydrodynamic pore radius r_h measures the distance from the pore center to the radius where the streaming velocity reaches zero. In this representation the sticking layer boundary condition is indicated by a negative slip length b whereas a positive slip length is typical of a slip boundary condition. The standard no-slip

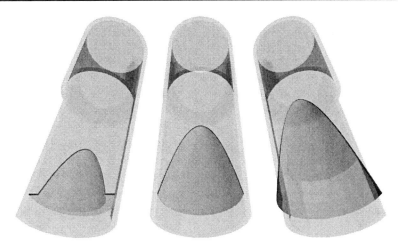

Figure 2.2.: Illustration of the possible boundary conditions along with the corresponding parabolic velocity profiles in a cylindrical tube with radius r_0. Mass transport takes place only where the streaming velocity is different from zero. (left): The reduction of the net flow rate is due to sticking layers at the pore walls, which do not participate in the mass transport. In addition the maximum velocity in the pore center is smaller than for no-slip boundary conditions (middle) because of the smaller hydrodynamic pore radius $r_h < r_0$. This gives rise to a further dramatic decrease in the flow rate. (right): In contrast, a slipping liquid with a hydrodynamic pore radius $r_h > r_0$ causes the highest streaming velocity and consequently the highest net flow rate.

condition yields $b = 0$ meaning $r_0 = r_h$.

It is obvious that such modified boundary conditions and thereby altered flow rates play a crucial role in systems that are highly dominated by fluid-substrate interfaces. This applies for extremely miniaturized systems such as lab-on-a-chip applications, in which fluid amounts of some picoliters only are manipulated. The enormous academic and economic interests on the interfacial behavior of liquids are manifested by a vast publication rate concerning this issue during the last decade. Many different techniques like SFA, atomic force microscopy (AFM), particle image velocimetry (PIV), fluorescence recovery after photobleaching (FRAP) and controlled dewetting as well as molecular dynamics (MD) or lattice Boltzmann simulations were utilized.

Up to date many factors have been found that seem to influence the boundary conditions. The most prominent and maybe the least controversially discussed amongst them is the fluid-wall interaction expressed in terms of the wettability [11–19]. The weaker the interaction is the more likely is slip. In addition, shear rates beyond a critical value are supposed to induce slip, too [20–23]. In contrast, the influence of surface roughness is rather debatable [24]. There are results for a decrease [12, 25] as well as for an increase [26] of the slip length with increasing surface roughness. Furthermore, dissolved gases [27, 28], the shape of the fluid molecules [16] or the add-on of surfactants [29] seem to influence the boundary conditions. To sum up, there is a huge set of factors (see Refs. [27, 30, 31] for some good reviews) and certainly a complex interplay between many of them finally determines the interfacial behavior.

Because of the modified boundary conditions one has to substitute r_h for r in Eq. (2.1). This procedure yields

$$K = \frac{\phi_0}{8\tau} \frac{r_h^4}{r_0^2} = \frac{\phi_0}{8\tau} \frac{(r_0 + b)^4}{r_0^2} \tag{2.7}$$

for the permeability of the membrane. Equation (2.7) illustrates the high sensitivity of K on b, provided b is on the order of or even larger than r_0. Therefore, measuring the hydraulic permeability gives direct access to the slip length b for a given liquid under given conditions.

One has to keep in mind that boundary conditions and fluid properties derived from measured flow rates are subject to a central restriction: one cannot verify the predefined parabolic shape of the velocity profile in the mesoscopic flow geometry. This is because there is no direct access to the profile itself but only to flow rates, which correspond to the velocity profile integrated over the whole pore cross-sectional area. Nevertheless, molecular dynamics simulations prove the formation of parabolic flow profiles even down to channel radii of 3 molecular diameters [32–34] and, hence, justify inferences based on this major assumption.

3. MATRIX PROPERTIES

This chapter will give some information on the fabrication and on the most important properties of the porous matrices that were applied in the measurements for this thesis. These samples can be devided into two classes: on the one hand there is porous Vycor® glass as the classic representative for an isotropic (sponge-like) network. On the other hand there are the anisotropic (sieve-like) matrices of porous silicon with aligned pores. Details on their characterization by means of volumetric sorption isotherms and gas permeation measurements will be given in Appendix B.

3.1. Porous Vycor® Glass

Most measurements on which this thesis is based, were performed with porous Vycor® (code 7930) provided by Corning Incorporated (see Fig. 3.1 for a raytracing illustration). It is produced through metastable phase separation in the alkali-borosilicate-glass system SiO_2-B_2O_3-Na_2O (known as spinodal decomposition). The following extraction of the alkali-rich phase leads to a sponge-like pore network with pores on the nanometer scale embedded in a matrix mostly consisting of SiO_2 (96.3 %) and a small amount of B_2O_3 (3 %) [35]. The monolithic Vycor® is featured by its easy availability as well as its transparency and mechanical robustness, which allow for an easy handling and shaping of the delivered glass rods and their utilization in optical experiments. However, the production process entails a rather broad pore size distribution $P(r)$, which will be addressed in the following.

Figure 3.1.: Raytracing illustration of the sponge-like internal structure of a porous Vycor® cuboid with its tortuous pore network.

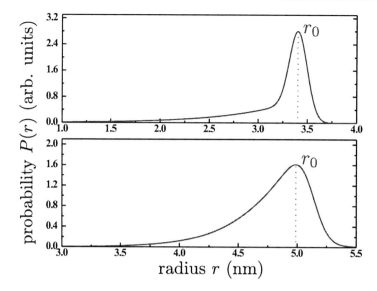

Figure 3.2.: Distribution of pore radii as extracted from nitrogen sorption isotherm measurements for both V5 (upper panel) and V10 (lower panel). The peak values of the mean pore radii r_0 (3.4 nm and 4.9 nm) are indicated by the dotted lines. For more details see Appendix B.

Two different batches of Vycor® glass were applied that differed in the mean pore radius r_0 whereas they coincided in the volume porosity $\phi_0 \approx 0.3$. For convenience I will refer to them as V5 ($r_0 = 3.4$ nm) and V10 ($r_0 = 4.9$ nm) from now on.

The matrix properties were accurately ascertained employing nitrogen sorption isotherms conducted at 77 K (see upper panel in Fig. B.2) and through a subsequent analysis within a mean field model for capillary condensation in mesopores. Finally, one arrives at the pore size distributions $P(r)$ shown in Fig. 3.2. Both batches are characterized by a most probable pore radius (the mean pore radius r_0) and an asymmetric distribution of several larger but of far more smaller pores. This particular shape will play an important role over the course of the investigation of the invasion kinetics in porous Vycor® being discussed in chapter 7.

In the meantime it is sufficient to characterize the samples by their mean pore

Table 3.1.: Properties of the two Vycor® batches as extracted from measurements presented in Appendix B.

sample batch	mean pore radius r_0	volume porosity ϕ_0	tortuosity τ
V5	(3.4 ± 0.1) nm	0.3 ± 0.005	3.6 ± 0.4
V10	(4.9 ± 0.1) nm	0.315 ± 0.005	3.6 ± 0.4

radii r_0 neglecting the distributions. Interestingly, both measured radii disagree with information from the official data sheet ($r_0 = 2.0$ nm) but the value for V5 agrees with preceding studies, which provide $r_0 = 3.4$ nm from sorption isotherm measurements [36, 37] and $r_0 \approx 3.8$ nm from a standard analysis of transmission electron micrographs [37]. Conversely, there are no such disagreements concerning its porosity. The data sheet value $\phi_0 = 0.28$ coincides with the measured ones $\phi_0 \approx 0.3$ for V5 and $\phi_0 \approx 0.315$ for V10. The same is true for the evaluation of the transmission electron micrographs yielding $\phi_0 = 0.31$ [37].

In order to gain information on the samples' tortuosity in-house gas permeation experiments [2] were carried out. They revealed no differences between V5 and V10 (see Appendix B). In agreement with SANS measurements [1] and simulations of the glass' internal topology [3] a value of $\tau \approx 3.6$ was found for both batches. The properties are summarized in Tab. 3.1.

The Vycor® membranes are highly hydrophilic. This is a consequence of glass being a high-energy surface with chemical binding energies on the order of 1 eV. Nearly any liquid spreads on such surfaces. This behavior can be comprehended considering the Young-Dupré equation (with the indices *Solid*, *Liquid* and *Vapor* of the interfacial tension γ and the static contact angle θ_0)

$$\gamma_{SV} = \gamma_{SL} + \gamma_{LV} \cos \theta_0 . \tag{3.1}$$

The empirical Zisman criterion predicts that any liquid fulfilling $\gamma_{LV} < \gamma_C$ (with the critical surface tension γ_C of the surface) totally wets this surface. For glass it is $\gamma_C \approx 150 \frac{mN}{m}$ [38]. Hence, even highly polarizable liquids like water spread on silica surfaces (meaning $\theta_0 = 0°$). This assumption was also tested experimentally and proved true for all liquids taken into consideration for this thesis by means of contact angle measurements in the group of Karin Jacobs[1]. It is exemplarily illustrated for water on surface-oxidized silicon in Fig. 3.4 (middle).

What is more, silica substrates provide the simple opportunity to alter the surface chemistry and thereby reduce the surface energy. This can be done according to the process illustrated in Fig. 3.3. As a consequence the wettability of the pore walls can crucially be modified. This is expressed by the critical surface tension being decreased down to $\gamma_C \approx 20 \frac{mN}{m}$ [38]. Water, for example, usually does not wet such treated surfaces anymore. This behavior is illustrated in Fig. 3.4 (right). Through such modifications studies of the influence of the liquid-substrate interaction on the dynamics become feasible.

3.2. Porous Silicon

Pore networks like the monolithic Vycor® glass just presented certainly have many advantages. From their mechanical robustness and optical transparency to their easy availability there are many facts that make them particularly suitable for the imbibition measurements performed for this thesis. For some purposes isotropic

---

[1] Soft Condensed Matter Physics Group at Saarland University, Saarbrücken, Germany.

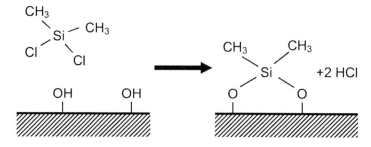

Figure 3.3.: Illustration of the reduction of the surface energy through silanization. In the presence of dimethyldichlorosilane $(Si(CH_3)_2Cl_2)$ the polar and consequently high-energy hydroxyl (OH) groups at the glass surface are replaced by low-energy methyl (CH_3) groups.

networks are not applicable, though. This becomes apparent when dynamic or structural properties of adsorbents should be studied with respect to their orientational dependency in the aniostropic pore geometry. Such measurements are typically carried out applying neutron, X-ray or light scattering techniques and do always require straight and aligned pores (alternatively in words of the tortuosity: $\tau = 1$).

Some supplemental experiments presented within this thesis were conducted with porous silicon membranes (see Fig. 8.6, Fig. 8.13, and Fig. B.7). These matrices are produced in-house employing an electrochemical anodic etching process of Si (100) wafer according to a standard recipe [39, 40]. The pores are linear, non-interconnected and oriented along the $\langle 100 \rangle$ Si crystallographic direction (perpendicular to the membrane surface).

Depending on the current density i, the hydrofluoric acid (HF) concentration c, the doping of the wafer and the etching time the membrane parameters can be tuned within certain limits. By default highly p-doped wafers with a resistivity of $0.01 - 0.02\,\Omega\,cm$ were used. Given the standard values ($i = 12.5\,\frac{mA}{cm^2}$ and $c = 20\,\%$) this generates membranes with a mean pore radius $r_0 \approx 6\,nm$ and a porosity $\phi_0 \approx 0.5$. The etching depth and accordingly the pore length can be controlled via the etching time. As a rule of thumb the etching front advances $40 - 50\,\mu m$ per hour. On demand, the porous matrix can be detached from the underlying

Table 3.2.: Properties of in-house produced porous silicon as extracted from measurements presented in Appendix B. The thickness of the membrane can be controlled via the etching time.

property	value
mean pore radius	6 nm
porosity	0.5
tortuosity	1

Figure 3.4.: Illustration of the impact of the surface chemistry on the wettability of a substrate elucidated by photographs of water droplets on a planar surface-modified silicon substrate. (left): Silicon as prepared: $\theta_0 \approx 50°$. (middle): Silicon after 23 h chemical oxidation in H_2O_2: $\theta_0 < 10°$. (right): Silicon after 23 h chemical oxidation in H_2O_2 and subsequent silanization in $Si(CH_3)_2Cl_2$: $\theta_0 \approx 90°$. Courtesy of Anke Henschel, Saarland University, Saarbrücken, Germany.

silicon wafer after finishing the etching process. The discussed properties are summarized in Tab. 3.2.

The pore walls of the obtained matrices carry silane groups (Si−H) and thus are not hydrophilic (see Fig. 3.4 (left)). Without further treatment only low-energy liquids enter the membrane. Fortunately, the surface chemistry can easily be changed. Oxidizing the surface silane to silanol (Si−O−H) by using hydrogen peroxide (H_2O_2) enlarges the surface energy and hence renders the membrane hydrophilic (see Fig. 3.4 (middle)). One can even go further and substitute a thermal oxidation process for the chemical one. Because of the tiny interpore distances of only a few nanometers the diffusion length of oxygen at moderate temperatures (800 °C) is sufficient to oxidize not only the surface but the whole sample. The resultant totally transparent silica membranes (see Fig. 3.5) are then applicable in optical measurements as well [41].

The microscopic structure of porous silicon is depicted by the transmission electron micrographs in Fig. 3.6. The mean pore radius mentioned above and extracted from sorption isotherms is in high accordance with systematic analyses of TEM images. A remarkable feature is the extraordinary shape of the pore's circumference. The apparent faceting is induced by the underlying crystalline structure of the silicon (see Fig. B.4). These irregularities give rise to some unexpected effects concerning both the static structure [42] and the dynamics [43] of an adsorbate.

However, the irregular pore geometry can be modified to a more smooth and tubular one. This can be done by controlled partial oxidation of the pore walls (using hydrogen peroxide) and subsequent dissolution of the resultant silica corona (using hydrofluoric acid) [44]. As an additional result of this precedure the mean pore diameter is enlarged and the volume porosity is increased. Therefore this treatment is also suitable if the standard pore diameter is too small for the desired purposes.

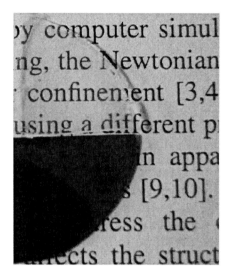

Figure 3.5.: Detached porous silicon membrane as prepared (bottom) and thermally oxidized to silica (top). Courtesy of Patrick Huber, Saarland University, Saarbrücken, Germany.

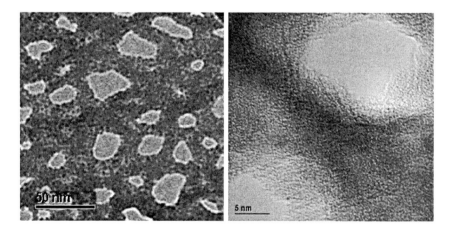

Figure 3.6.: Transmission electron micrographs of porous silicon with different magnification factors. The sample was thinned by means of Ar ion milling in order to ensure a sufficient electron transmission rate. The pores correspond to the bright areas. Courtesy of Jörg Schmauch, Saarland University, Saarbrücken, Germany.

Part II.

Spontaneous Imbibition Study

Part II of this thesis includes a crucial study of the dynamics of mesopore-confined liquids ranging from water to hydrocarbons to silicon oils and liquid crystals. For this purpose I utilized the capillary rise of a wetting liquid in a porous substrate also known as spontaneous imbibition. The basic principles and experimental methods will be introduced prior to presentation of the measurements and their interpretation.

4. Basic Knowledge

Spontaneous imbibition of a liquid in porous material is a very common phenomenon. From everyday life almost everybody knows the effect of coffee or tea being absorbed by a cube of sugar. A rather unpleasant instance of being confronted with its consequences is rising moisture in your basement walls. Another well-known example is the transport of water from the roots of a tree up to its limbs, branches and leaves, which is partly driven by capillary action [45].

4.1. Basic Principles

From the physisist's point of view spontaneous imbibition is an impressive example for interfacial physics. The driving force behind the capillary rise process is the Laplace pressure p_L acting on the curved meniscus of a liquid in a pore or porous structure. The examples given above represent a small selection of the addressed topologies such as compressed sugar grains, cracks and pores in masonry blocks or concrete or the xylem network in trees. And finally these structures are also represented in the tortuous pore network in the porous Vycor® glass, which is used in this thesis for examinations of the flow dynamics of liquids in spatial confinement.

The Laplace pressure is specified by

$$p_L = \sigma \left(\frac{1}{R_1} + \frac{1}{R_2} \right) \tag{4.1}$$

with σ being the surface tension of the liquid and R_1 and R_2 being the local radii of curvature of the liquid surface. For that reason it is important to obtain information on the shape of the meniscus. In general it is determined by an interplay of surface forces and gravity. The influence of the latter can be estimated by means of the so-called capillary length

$$\lambda_c = \sqrt{\frac{\sigma}{\rho g}} \tag{4.2}$$

with g being the acceleration due to gravity and ρ representing the density of the liquid. For water (with $\sigma \approx 72 \frac{mN}{m}$ and $\rho = 1 \frac{g}{cm^3}$) Eq. (4.2) yields a capillary length of 2.7 mm. For surfaces with characteristic lengths smaller than λ_c the gravitational force can be neglected. Therefore, in the porous structure of Vycor® with mean pore diameters on the order of 10 nm the menisci are solely determined by surface forces. Consequently, they take the shape of a spherical cap (see Fig. 4.1) described by a unitary radius of curvature $R_1 = R_2 \equiv r_c$ resulting in a uniform Laplace pressure $p_L = \frac{2\sigma}{r_c}$ at each point of the meniscus.

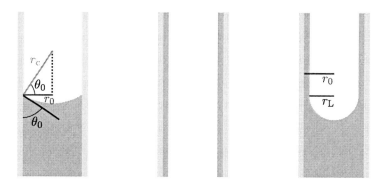

Figure 4.1.: Sketches of cuts through pores before and during an imbibition experiment. (left): Spherical meniscus of a liquid in a pore with the static contact angle θ_0. From simple geometric considerations one can deduce the relation between the radius of curvature r_c and the pore radius r_0: $r_c = \frac{r_0}{\cos\theta_0}$. (middle): Water coating of the silica pore walls because of the finite humidity in the laboratory. (right): Illustration of the difference between the pore radius r_0 and the Laplace radius r_L due to preadsorbed water layers, meaning that $r_c = \frac{r_L}{\cos\theta_0}$.

Eventually, from simple geometric considerations, one can deduce the relation between the radius of curvature r_c and the pore radius r_0 (see Fig. 4.1 (left)) yielding

$$p_L = \frac{2\sigma\cos\theta_0}{r_0} \tag{4.3}$$

for the Laplace pressure. It is obvious that the interaction between liquid and matrix expressed in terms of the static contact angle θ_0 plays a crucial role in the capillary rise process. Spontaneous imbibition can only occur for a wetting or partially wetting liquid that is a liquid whose contact angle with the matrix is smaller than 90°. The influence of the contact angle hysteresis and a possible substitution of the dynamic contact angle θ_D for θ_0 will be discussed in section 4.6.

Of course, as the liquid rises beyond its bulk reservoir to a certain level h the hydrostatic pressure $p_h = \rho g h$ acting on the liquid column increases. Accordingly the driving pressure Δp must be modified to $\Delta p = p_L - p_h$. The final state shown in Fig. 4.2 is then derived from a balance between p_L and p_h. The maximum rise level h_{max} is given by Jurin's law:

$$h_{max} = \frac{2\sigma\cos\theta_0}{\rho g r_0} . \tag{4.4}$$

It is worthwhile calculating h_{max} for water in the silica network of Vycor®. Assuming $\theta_0 = 0°$ for water on a glass substrate and $r_0 \approx 5\,nm$ Eq. (4.4) yields $h_{max} \approx 2.9\,km$ corresponding to a Laplace pressure of 290 bar. Thus, for the here examined rise levels restricted by the maximum sample height to less than 5 cm the gravitational force can be neglected meaning that

$$\Delta p = p_L . \tag{4.5}$$

Figure 4.2.: Final state of the capillary rise process of dyed water in three glass capillaries with different diameters. Applying Jurin's law Eq. (4.4) one can deduce the capillary's diameter r_0 from the maximum rise level h_{max}. Courtesy of Hans Peter Läser, Soil Physics, Institute of Terrestrial Ecology, ETH Zürich, Swiss.

The prevalence of surface forces over gravitation can likewise be expressed in terms of the dimensionless Bond number Bo:

$$\mathrm{Bo} \equiv \frac{\text{gravitational force}}{\text{capillary force}} = \frac{\rho g \mathcal{L}^2}{\sigma} \ll 1 \qquad (4.6)$$

because of the tiny characteristic length scales $\mathcal{L}$.

4.2. Influence of Preadsorbed Liquid Layers

In order to gain information on the dynamics of the imbibition process one needs to recall Darcy's law Eq. (2.4). Considering Fig. 4.3 (left) one is able to conclude that the sample height d appearing in Darcy's law has to be replaced by the actual rise level $h(t)$ since only the filled parts of the sample contribute to the flow dynamics. Moreover, at any given time t the imbibed fluid volume $V(t)$ is closely related to $h(t)$ via

$$V(t) = \phi_{\mathrm{i}} A h(t) . \qquad (4.7)$$

Here ϕ_{i} denotes the initial volume porosity of the Vycor® sample. This porosity is reduced with respect to ϕ_0 because of water layers on the silica pore walls, which are immediately adsorbed under standard laboratory conditions because of the finite humidity and the highly attractive interaction between water and silica (see Fig. 4.1 (middle)). This wall coating amounts to a 15 % to 20 % decrease in porosity. The adsorbed water is highly stabilized and can only be removed at elevated temperatures [46]. Especially the physisorbed first layer requires evacuation at temperatures $T > 150\,^{\circ}\mathrm{C}$ for removal.

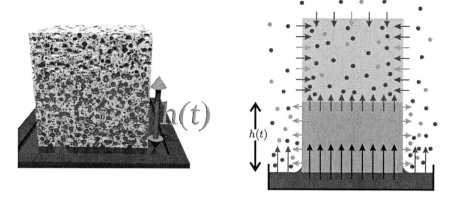

Figure 4.3.: (left): Raytracing illustration of a Vycor® sample during a capillary rise experiment filled up to the level $h(t)$. (right): Illustration of the evaporation processes superimposing the imbibition measurement (black arrows). From both the liquid reservoir (blue arrows) and the already filled (green) parts of the sample (green arrows) liquid evaporates from the system (blue and green spheres) and might afterwards condense in the still empty (yellow) parts of the sample. Furthermore, liquid vapor from the advancing imbibition front might as well directly invade the empty parts beyond (red arrows and spheres).

The exact degree of coating depends on the sample temperature and the absolute humidity, thus it cannot be known beforehand. Fortunately ϕ_i can be extracted for each sample from the performed mass increase measurements, which will be presented later (see Fig. 6.1). Given the density ρ of the imbibed liquid and the volume V_s of the sample block one gains direct access to the initial porosity via the overall mass increase M due to the liquid uptake (indicated by the blue arrow in Fig. 6.1):

$$\phi_i = \frac{M}{\rho V_s} \, . \tag{4.8}$$

Similarly, the Laplace pressure must be influenced by the initial wall coating. The preadsorbed water layers necessarily lead to a reduction of the radius of curvature of the menisci (see Fig. 4.1 (right)). This effect will be taken into consideration by substituting r_L for r_0 in Eq. (4.3) with $r_L \leq r_0$, meaning

$$\Delta p = p_L = \frac{2\sigma \cos\theta_0}{r_L} \, . \tag{4.9}$$

I do not have any reliable information on the layer thickness $x = r_0 - r_L$ and hence on the Laplace radius r_L. It can only be estimated from the difference between ϕ_0 and ϕ_i assuming the pores to be perfect cylinders with a radius r_0. Then $r_L = r_0 \sqrt{\phi_i/\phi_0}$ enables one to estimate the upper bound $x \leq 5\,\text{Å}$ (depending on temperature and humidity). Based on this result I will assume $r_L = (r_0 - 2.5\,\text{Å}) \pm 2.5\,\text{Å}$ in order to account for this effect. Fortunately, the impact of r_L on the overall imbibition dynamics is comparatively small (see Eq. (4.14)) and, therefore, the uncertainty in the actual Laplace radius has only little effect on the final results.

4.3. Thermodynamics of the Advancing Menisci

So far I have assumed the capillary rise process to be the predominating effect responsible for the occurring liquid dynamics. However, they could also occur because of capillary condensation processes via the vapor phase – the effect utilized in sorption isotherm measurements. Recently conducted lattice Boltzmann simulations predict a breakdown of the usual imbibition dynamics caused by such condensation processes [47]. For that reason it is worth considering its impact on the measurements.

Capillary condensation can happen through liquid evaporation from either the advancing liquid front in the sample or the bulk liquid reservoir beneath (see Fig. 4.3 (right) for an illustration of various processes). A by-passing via the latter mechanism (blue arrows) can easily be prevented by sealing the top and side facets of the sample. Especially for volatile liquids like water or short-length hydrocarbons such a sealing is of the highest importance not only because of the material influx but likewise because of the liquid evaporation from the sample (green arrows). In the discussion of the sample preparation this point will be reconsidered.

Evaporation from and subsequent condensation beyond the advancing liquid front (red arrows) cannot be prevented by such measures, though. In order to estimate its influence on the overall dynamics one has to consult the results from the sorption isotherms presented in Appendix B. According to these measurements the reduced vapor pressure P beyond the curved menisci only has the ability to generate filling degrees between 20 % and 30 % in adsorption. This corresponds to maximum three layers and is not sufficient to induce noticeable capillary condensation [48]. Consequently, one may conclude that evaporation from the advancing front at the most excites filling degrees of 30 %. But this value has to markedly be reduced due to the already existent initial coating of the pore walls discussed in section 4.2.

In summary it can be said that a by-passing material transport via the vapor phase can occur, if only to a rather limited extent. Eventually, neutron radiography and optical measurements will not only verify this prediction, but also permit more quantitative statements on the impact of evaporation and condensation mechanisms on the overall imbibition dynamics.

4.4. Dynamics of the Imbibition Process

With all these preliminary considerations in mind I will now return to the dynamics of the imbibition process. Darcy's law Eq. (2.4) in conjunction with Eq. (4.7) leads to the simple differential equation

$$\dot{h}(t)\,h(t) = \frac{K}{\phi_i\,\eta}\Delta p \qquad (4.10)$$

solved through

$$h(t) = \sqrt{\frac{2\,K}{\phi_i\,\eta}\,\Delta p}\,\sqrt{t}\,. \tag{4.11}$$

Nowadays this square root of time behavior is often referred to as the Lucas-Washburn (LW) law after its alleged discoverers [49, 50]. Actually, the $\sqrt{t}$-scaling was first found by Bell and Cameron [51] more than 10 years earlier. Following a proposal by Howard Stone [52] I will term Eq. (4.11) BCLW law in order to acknowledge all contributors.

Given the driving pressure Δp in accordance to Eq. (4.9) and the permeability K corresponding to Eq. (2.7) one obtains the rise level h at a certain time t

$$h(t) = \underbrace{\sqrt{\frac{\sigma\,\cos\theta_0}{2\,\phi_i\,\eta}}\,\Gamma}_{C_h}\,\sqrt{t}\,. \tag{4.12}$$

Along with Eq. (4.7) and the density ρ of the liquid Eq. (4.12) results in

$$m(t) = \underbrace{\rho\,A\,\sqrt{\frac{\phi_i\,\sigma\,\cos\theta_0}{2\,\eta}}\,\Gamma}_{C_m}\,\sqrt{t}\,, \tag{4.13}$$

the mass increase m of the sample due to the liquid uptake as a function of the time t. The so-called imbibition strength Γ in Eqs. (4.12) and (4.13) is given through

$$\Gamma = \frac{r_h^2}{r_0}\,\sqrt{\frac{\phi_0}{r_L\,\tau}}\,. \tag{4.14}$$

The corresponding flow configuration is illustrated in Fig. 4.4. Comparable to the hydraulic permeability Eq. (2.7) the imbibition strength depends solely on the matrix' internal structure. All liquid and temperature specific quantities

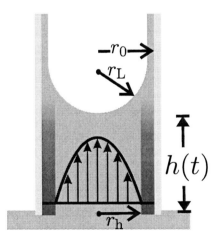

Figure 4.4.: Schematic view of the capillary rise in a cylindrical pore with pore radius r_0, Laplace radius r_L and hydrodynamic pore radius r_h.

and the sample's shape do not influence Γ, thus it should be a constant for all measurements with the same sample type.

Interestingly, relation (4.14) reveals that in principle the capillary rise dynamics are proportional to the square root of the capillary's radius: $\Gamma \propto \sqrt{r}$. As a consequence, in common capillary rise experiments with channel diameters on the order of some hundred micron the rise dynamics are so fast that without any additional instruments merely the steady-state equilibrium configuration at the very end of the process can be observed (see Fig. 4.2). This is contrary to imbibition in mesoporous networks. Here, the much smaller pores induce a lot more viscous drag, which reduces the overall dynamics significantly. Typical rise times in my experiments are on the order of some hours or even days. This fact allows for an easy recording of the dynamics by measuring either the sample's mass increase $m(t)$ or the rise level $h(t)$ as a function of the time.

4.5. Light Scattering at the Invading Interface

Commonly the imbibition dynamics are studied via a recording of the sample's mass increase due to liquid uptake. For one thing such measurements are rather easy to perform. The corresponding setup depicted in Fig. 5.1 suggests a metaphor of physics using kitchen scales and a wrist watch. For another thing measuring the rise level $h(t)$ is not feasible at all. Most likely this is a bit confusing since the phenomenon of the meniscus (and the liquid column) rising above its reservoir is normally the most outstanding manifestation of the capillary rise process (see Fig. 4.2). Correspondingly, the observation of the advancing menisci with a camera at first glance appears as the most obvious way for measuring $h(t)$.

Unfortunately, for the transparent liquids regarded here (water, hydrocarbons, and silicon oils) there is no optical difference between the filled and the empty sample, provided the liquid is not artificially dyed (e.g., ink). What is more, light scattering is also not observable since one of its two necessary preconditions is not fulfilled. Admittedly, there is a difference in the refractive index between the silica substrate and the air in the empty or the liquid in the filled pores. But the spatial variations of the refractive index roughly equivalent to the pore-pore distance are on the order of 25 nm only [53], thus being too small to induce any scattering of visible light.

Nevertheless, during the measurement one is able to observe a white front wandering from the bottom of the sample to its top where it finally vanishes (see first row in Fig. 4.5 and Fig. 7.2). This phenomenon can very well be attributed to a special arrangement of the liquid in this region. It is commonly known from the drainage of completely filled porous samples with tortuous networks such as Vycor® [54–56]. The process of drying is never homogenous. The sample rather empties via percolating paths meaning that there are connected regions of already empty pores alternating with still filled ones. Now the crucial point is that these regions vary partly on length scales comparable with the wavelength of visible light and hence imply significant light scattering. The whole sample turns white

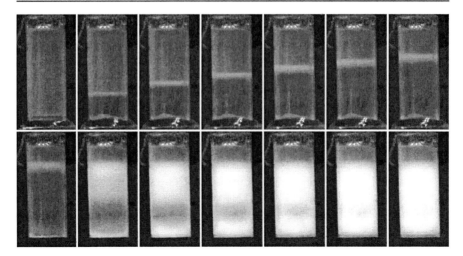

Figure 4.5.: Series of pictures of a detachment experiment of water invading a porous Vycor® block (V10, $V_s = 6.75 \times 4.85 \times 16.70\,\mathrm{mm}^3$) at room temperature. Top and side facets were sealed. (first row): Standard imbibition with white advancing front. Duration between each picture: 10 min. (second row): Detachment from the reservoir and turning white of the filled parts. Intervals between pictures: 30 min.

(see second row in Fig. 4.5 and Refs. [54–56]).

Consequently, the white front appearing during the capillary rise experiment includes information on the arrangement of the liquid in this region. Its mere appearance indicates an additional process taking place at the advancing front. There is no sharp boundary between filled and empty parts and as a result there seems to be no such thing as *the* rise level. Shedding light upon the observed roughening of the advancing front will be part of this thesis as well. For that reason some rise experiments were recorded by means of a CCD camera. The dynamics of the invading interface will be addressed in chapter 7.

4.6. Remarks on Short Time Effects

Finally, I will outline some effects that dominate the imbibition process in its very initial phase, meaning the first few nanoseconds. However, these effects occur on time scales far beyond the temporal resolution limit of the experiment (on the order of one second) and therefore do not play a decisive role for the measurements, which usually take several hours or even days. But these phenomenons should not remain unmentioned.

4.6.1. BCLW Law in Mesopores

To date the capillary rise behavior in meso- and nanopores has several times been studied by means of molecular dynamics simulations. At first glance the results are rather ambiguous. Some examinations prove the $\sqrt{t}$-law to be true [34, 57] down to pores with radii less than 1 nm. Nonetheless, a series of publications by Quirke *et al.* on imbibition in carbon nanotubes (CNT) suggests a linear function of the time [58, 59]. They attributed this behavior to the atomic smoothness of the nanotubes. But one has to keep in mind that in their simulated experiments it takes only 82 ps until the tube is completely filled. Interestingly Binder *et al.* found a similar behavior restricted to a comparable short transient regime at the very beginning of the simulation followed by the classic BCLW law [34].

Anyhow, these results can uniformly be explained. The BCLW law neglects inertial effects, thus it is only valid for time scales sufficient to establish viscous flow. For very short times a more general ansatz has to be applied (Bosanquet equation), which approaches the BCLW law for long times. The observed deviations from the BCLW law are possibly caused by a high degree of turbulences and initial deformations of the velocity profiles. This lasts until the flow reaches its typical low Reynolds numbers, for which turbulent flow cannot play a role at all.

The typical time τ_{init} that is required for the viscous flow to establish itself in the pore can be estimated with [38]

$$\tau_{init} \approx \frac{\rho r_0^2}{\eta} \,. \tag{4.15}$$

This typically yields times on the order of some 10 ps in high agreement with the MD studies just presented.

According to Ref. [60] the fluid-wall interaction, and in particular the surface friction due to molecular corrugation, additionally seem to influence τ_{init}. For the atomic smooth CNT τ_{init} is rather increased and a lengthy non-BCLW behavior can be observed. The nearly frictionless surface of the nanotubes is also supposed to be responsible for the enhanced mobility of water and hydrocarbons flowing through these tiny ducts [61, 62]. Nevertheless, this phenomenon occurs on time scales far beyond the temporal resolution limit of the experiment and does not affect my measurements at all [38, 63].

4.6.2. Dynamic Contact Angle

The static contact angle θ_0 of a drop resting on a surface is a constant determined by an interplay of the interactions between the solid, the liquid and the vapor phase. As a rule this assumption fails if the contact line begins to move. Here, a different contact angle, the so-called dynamic contact angle θ_D (see Fig. 4.6) is observed. It lies somewhere between the advancing (the largest achievable) and the receding (the smallest achievable) contact angle. Since θ can be increased or decreased (with respect to θ_0) depending on whether the contact line begins to

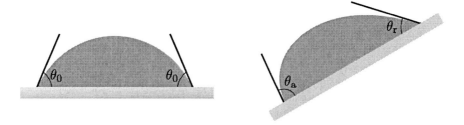

Figure 4.6.: Illustration of the dynamic contact angle. In the static state (left) on both sides of the drop the static contact angle θ_0 is established. Though, on an inclined plane (right) the drop's shape becomes asymmetric with the dynamic contact angles $\theta_r < \theta_0$ and $\theta_a > \theta_0$ on the receding and the advancing side, respectively.

move in the direction of the gas phase or in the direction of the liquid phase this phenomenon is often referred to as contact angle hysteresis [38, 64].

It is this dynamic contact angle that is required as a boundary condition for modeling problems in capillary hydrodynamics. In consequence, it is important to estimate the changes that arise from this effect. Unlike the static contact angle, θ_D is not a material property. Actually, for small spreading velocities v expressed in terms of the capillary number Ca

$$\mathrm{Ca} \equiv \frac{\text{viscous force}}{\text{capillary force}} = v\,\frac{\eta}{\sigma}\,, \tag{4.16}$$

it seems to solely be influenced by the capillary number itself [65]. This is the statement of the most widespread work relation describing the dynamic contact angle, namely the Hoffman-Voinov-Tanner law $\theta_D^3 \propto \mathrm{Ca}$. It is valid for $\mathrm{Ca} < 10^{-4}$ [63] or, with some correction, for $\mathrm{Ca} < 10^{-2}$ [66].

Interestingly, Richard Hofmann carried out a systematic study of dynamic contact angles in glass capillary tubes for a wide range of capillary numbers $(4 \cdot 10^{-5} < \mathrm{Ca} < 36)$ [67]. Some years later Jiang *et al.* gave an empirical correlation for his results [68]:

$$\frac{\cos\theta_0 - \cos\theta_D}{\cos\theta_0 + 1} = \tanh\left(4.96 \cdot \mathrm{Ca}^{0.702}\right). \tag{4.17}$$

This expression can be used to estimate the impact of the contact angle hysteresis on the capillary rise experiments presented within this thesis. For this purpose one needs information on the prevailing capillary numbers being tantamount to the advancement speed v of the liquid front. One can deduce this speed as the time derivative of Eq. (4.12) resulting in

$$v(t) = \frac{\mathrm{d}h(t)}{\mathrm{d}t} = \frac{C_h}{2\sqrt{t}}. \tag{4.18}$$

The obvious divergence for $t \to 0\,\mathrm{s}$ is again the manifestation of the deficiency of the BCLW law for very short time scales and has already been thematized in the literature [38, 69].

The flow speed $v(t)$ after one second elapsed rise time (consistent with the temporal resolution limit of the experiment) is a plausible quantity for an upper estimate of the prevailing capillary numbers. Furthermore, the highest speed will be observed for water (because of its relatively high surface tension) in V10 (because of the relatively low viscous drag due to its larger pore radii). Along with the respective measuring results (that will be presented in chapter 6) I obtain as the absolutely highest capillary number occuring in my experiments

$$\mathrm{Ca_{max}} \approx 10^{-6} \,. \tag{4.19}$$

Accordingly, the above approximation for low Ca Eq. (4.17) can be applied. With the static contact angle $\theta_0 = 0°$ this finally yields the highest achievable dynamic contact angle

$$\theta_{\mathrm{D,max}} = 2.2° \tag{4.20}$$

with $\cos\theta_{\mathrm{D,max}} = 0.9993$ compared to $\cos\theta_0 = 1$. It is apparent that the phenomenon of the contact angle hysteresis needs no further consideration. This result agrees with MD studies of Martic *et al.* [70]. Indeed, they found an initial variation of θ_{D} for liquids sucked into channels with 5 nm radius. Nevertheless, it relaxed within some 10 ns towards θ_0 what is again far beyond the temporal resolution limit of the experiment.

What is more, the contact angle remaining $0°$ during the whole measurement entails an interesting side effect. In a recent lattice Boltzmann simulation Kusumaatmaja *et al.* [71] examined the influence of surface patterning on the capillary filling of microchannels. They found that for $\theta < 30°$ the structuring has only little effect on the filling process and for completely wetting liquids there is no observable influence at all. Therefore I do not have to worry about atomic irregularities of the pore shape.

5. EXPERIMENTAL SETUPS

In this chapter the measuring techniques that were applied in order to study the imbibition dynamics of liquids in porous Vycor® will be introduced. Both the experimental setups and the sample preparation for the respective method will be elucidated.

5.1. Gravimetric Measurements

As a standard, imbibition dynamics are studied via a recording of the samples' mass increase due to the liquid uptake $m(t)$ as a function of the time t. Primarily, this is done because of the rather simple implementation of such measurements. The corresponding setup is depicted in Fig. 5.1.

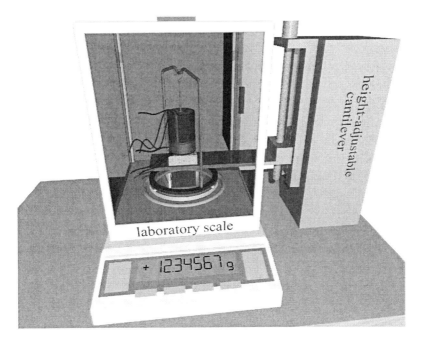

Figure 5.1.: Raytracing illustration of the liquid imbibition setup (LIS). The cell (see Fig. 5.2) is mounted on a height-adjustable cantilever reaching into the housing of a standard laboratory scale from Sartorius (model: BP211D).

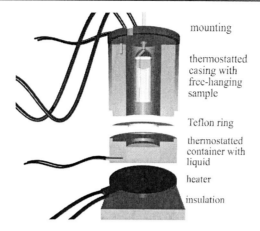

Figure 5.2.: Exploded view of the imbibition setup (raytracing illustration). The top of the Vycor® sample is glued to a wire and hangs freely in the cell. The cell itself consists of a casing and a container. Both are built of copper and can separately be thermostatted employing heating foils (red cables) and platinum resistance thermometers (Pt100, blue cables). The thermal decoupling of casing and container by means of a teflon ring effectuates an increase in temperature stability. Insulation against the metallic cantilever is established through a styrofoam block.

5.1.1. Setup

For a time dependent measurement of the force acting on and, hence, of the mass increase of the porous Vycor® block the sample is installed on a standard laboratory scale using a special mounting. Each second its mass can be logged into a text file.

In order to perform measurements beyond room temperature a cell was constructed (see Fig. 5.2) that allows for a simultaneous thermostatting of the sample itself and the liquid reservoir beneath by means of a LakeShore (model: 330) and a Eurotherm (model: 2416) temperature controller (TC), respectively. The Vycor® block hangs freely in this cell. To ensure high temperature accuracy the temperature T within the casing was measured employing a thermistor thermometer (HANNA, code: HI9040) and related to the TC monitoring T_{TC}:

$$T[°\mathrm{C}] = 1.099 \cdot T_{\mathrm{TC}}[°\mathrm{C}] - 2.699\,°\mathrm{C}\,. \tag{5.1}$$

The measurement can be started right after the thermalization of sample and liquid (ascertained as explained in subsection 6.1.1) by moving the whole cell upward until the sample touches the liquid surface.

5.1.2. Sample Preparation

All presented experiments were conducted with porous Vycor® glass. The samples were cut into regular shapes of either cylinders (V5) or cuboids (V10) of known dimensions (cross-sectional area A and height h_0). Prior to measurement they were stored in a desiccator to avoid the uptake of impurities from the surrounding air. They could be connected to the mounting employing a pliable wire, which was glued on top of each sample, thereby simultaneously sealing its topside (see Fig. 5.2). To impede any significant migration of the glue into the samples a two-component adhesive ('UHU plus sofort fest 2 min') with a pot life of 2 minutes only was used.

As already mentioned before, for highly volatile liquids such as water and the n-alkanes up to decane an additional sealing of the samples' side facets is of the utmost importance. Partly this is because of the possible influx of molecules via the vapor phase, which would enhance the overall mass uptake especially in the very beginning of the measurement. However, with increasing rise level another problem gains importance, namely the evaporation from the sample. This would actually lead to a pinning of the rise level h determined by a balance between the liquid supplying imbibition rate and the liquid detracting evaporation rate (see Fig. 5.3).

For water such a sealing can very well be realized with Scotch® tape. Additionally, this retains the samples' transparency, which is rather important for the optical measurements. Unfortunately, for the short-length n-alkanes this procedure is inappropriate since they dissolve the adhesive connection between tape

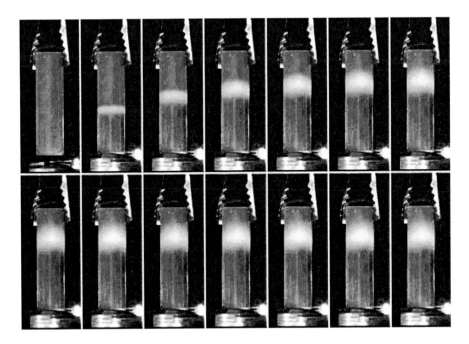

Figure 5.3.: Series of pictures of a pinning experiment of water invading a porous Vycor® block (V10, $V_s = 8.14 \times 6.05 \times 29.00\,\text{mm}^3$) at room temperature. The sample is not sealed. Intervals between pictures: 106 min. Due to the increasing impact of evaporation the front is eventually pinned.

and sample. For that reason the side facets of the block were also covered with the two-component adhesive mentioned above. For all other used liquids the samples' side facets remained unsealed.

5.2. Optical Measurements

As already explained in section 4.5 the capillary rise process is accompanied by the occurrence of a white front wandering from the bottom of the sample to its top where it finally vanishes. Meanwhile, its width gradually increases (see Fig. 4.5 and Fig. 7.2). This effect must be caused by a special arrangement of the liquid within this region. In order to shed light on this phenomenon some rise experiments were recorded optically.

5.2.1. Setup

For the observation of the front, pictures of the sample can be taken at regular intervals during the imbibition process. For this purpose a CCD monochrome camera (The Imaging Source, model: DMK 41BF02) can be installed in front of the housing of the laboratory scale. Of course, such measurements are only feasible if the copper casing for thermostatting is not required. This restriction limits the applicability of the method to liquids whose melting point is below room temperature.

5.2.2. Sample Preparation

All samples used were brought into rectangular shape in order to generate a comparable light scattering signal over the whole sample width. As to the rest, the sample preparation was identical to that of the gravimetric measurements.

5.3. Neutron Radiography Measurements

Neutron radiography is a particularly suitable method for gaining in-situ information on the space- and time-resolved distribution of the liquid within the sample (as opposed to the gravimetric measurements, which lack the spatial resolution). This enables one to study both the imbibition dynamics and the front roughening process in detail. Especially the light scattering phenomenon at the advancing interface can convincingly be explained only referring to the underlying liquid distribution. Furthermore, this method will reveal whether there is significant condensation of invading vapor beyond the liquid front or not. In section 4.3 such a by-passing material transport has only theoretically been excluded, provided appropriate measures are taken.

The advantages of neutron scattering as compared with X-ray techniques is impressively illustrated by Tab. 5.1. Since X-ray radiation is sensitive to the local

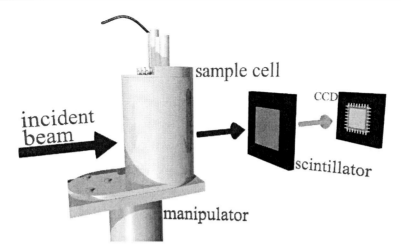

Figure 5.4.: Raytracing illustration of the structural principle of the neutron radiography setup ANTARES located at the research reactor FRM II of Technische Universität München in Garching, Germany. For a detailed view into the sample cell see Fig. 5.5. Blue arrows indicate neutron radiation, the green arrow indicates visible light.

electron density, its scattering cross section, and accordingly its attenuation, increases gradually with increasing atomic number. This results in a rather low contrast of hydrogen-, carbon- and oxygen-rich liquids (water, hydrocarbons) against the background of a silicon-rich matrix. This applies not for neutrons, for which the scattering cross section is not a function of the atomic number at all. With a neutron beam, a high contrast even between elements of low atomic

Table 5.1.: Microscopic total scattering cross sections σ_{tot} of thermal neutrons ($E = 25\,\mathrm{meV}$) and gamma radiation ($E = 100\,\mathrm{keV}$) for selected elements in their natural isotope composition. Data are taken from the ANTARES instrument homepage [72].

element	atomic number	σ_{tot} (barn) neutrons	σ_{tot} (barn) gamma rays
hydrogen	1	82	0.5
boron	5	770	2.5
carbon	6	5.6	3.0
oxygen	8	4.2	4.1
aluminum	13	1.7	7.6
silicon	14	2.3	8.6
gadolinium	64	50,000	810
lead	82	11	1,900

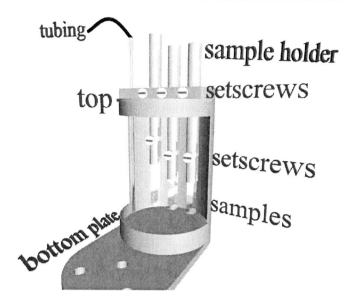

Figure 5.5.: Raytracing illustration of a view into the sample cell of the neutron radiography setup depicted in Fig. 5.4. The whole cell and the sample holder are made of aluminum. The samples are mounted on the clamp-like sample holder. Up to three of them can simultaneously be attached to the cell. The liquid is supplied via a tubing employing an external pump.

number can be obtained. Especially the high attenuation in hydrogen results in a large contrast in the Vycor® matrix. Despite its small amount of boron trioxide (B_2O_3) the latter shows only little neutron scattering.

5.3.1. Setup

The measurements were performed at the ANTARES (*A*dvanced *N*eutron *T*omography *A*nd *R*adiography *E*xperimental *S*ystem) beamline of the research reactor FRM II of Technische Universität München in Garching, Germany. The structural principle of ANTARES is illustrated in Fig. 5.4.

Neutrons are generated in the reactor core and moderated in a vessel filled with cooled liquid deuterium ($T = 25\,\text{K}$). The cold neutrons reach the experiment through a channel in the biological shielding and an adjacent evacuated (0.1 mbar) flight tube (length $\sim 12\,\text{m}$). Collimators with aspect ratio $L/D = 400$ mounted in the channels of the biological shielding create a parallel-aligned beam (maximum divergence of 0.3°). Subsequent apertures improve the resolution. An additional lead filter in the flight tube should reduce the unavoidable gamma radiation from the reactor core whereas it is almost transparent to neutrons (see Tab. 5.1). At the location of the specimen the neutron flux is approximately $10^8\,\text{s}^{-1}\text{cm}^{-2}$ and the illuminated area is $40.0 \times 40.0\,\text{cm}^2$.

The completely closed, cylindrical sample cell (see Fig. 5.5) and the sample holder are made of aluminum due to its weak interaction with neutron radiation. Up to three samples can simultaneously be attached to the setup. Via a tubing the initially empty cell can be filled with the liquid using an external precision pump, which can be switched on and off from beyond the experimental chamber. To ensure a simultaneous start of the capillary rise measurements of all mounted samples they must be brought to the same level prior to measurement. Thanks to the construction with separate sample holders the matrices can easily and independently be moved in vertical direction. Additionally, the whole sample cell can vertically and horizontally be moved employing a manipulator. This is important to make sure that all samples are well within the detector area (see dimensions of the CCD chip below).

For the space dispersive detection of the penetrated neutrons a gadolinium (Gd) detector (distance sample - detector $\sim 30\,\mathrm{mm}$) was used. Due to neutron capture Gd is activated and emits high energy gamma radiation, which ionizes a scintillator material (commonly zinc sulfide ZnS). The scintillator reemits the absorbed energy in the form of light flashes, typically in the visible range. Eventually, this signal can be recorded by means of a monochrome CCD camera. The dimensions of the CCD chip are $32.7 \times 32.7\,\mathrm{mm}^2$, its resolution is 2048×2048 pixels. This corresponds to a pixel height of $\sim 16\,\mathrm{\mu m}$ in the sample plane. The system resolution is on the order of 30 micrometers.

5.3.2. Sample Preparation

Again, all samples used were brought into rectangular shape with an identical cross-sectional area $(4.60 \times 4.60\,\mathrm{mm}^2)$ in order to generate comparable neutron transmission signals over the whole sample width and for all samples applied. Their heights were restricted to 10 mm, 15 mm, or 20 mm.

Rather problematic is the sealing of the samples' side facets. The standard Scotch® tape solution is inappropriate since such a tape would absorb a greater part of the incoming neutron intensity due to its composition of mostly organic compounds. For this reason I applied aluminum foil tape (3M, code series 591), which is almost transparent to neutron radiation. Apart from that, the top facets were again sealed by means of the two-component adhesive in order to guarantee a reliable tightness of the sealing.

6. Rise Dynamics in Porous Vycor®

The dynamics of the capillary rise in porous Vycor® give information on the mesoscopic flow behavior. This chapter represents a crucial study of this process for various liquids ranging from water to hydrocarbons and silicon oils. The results of the underlying experiments will eventually shed light on the detailed boundary conditions and the applicability of classical hydrodynamics for such restricted geometries. Potential influences of the dimensions of the liquids' building blocks on the flow dynamics will be investigated. Moreover, the influence of a by-passing material transport via the vapor phase will be assessed.

6.1. Applied Methods

The capillary rise dynamics were studied with two different methods. Principally, all liquids applied were investigated by means of gravimetric measurements. For certain representative liquids additional neutron radiography experiments were conducted. The measuring principles for the two methods are explained below.

6.1.1. Gravimetric Measurements

The measuring principle of the gravimetric experiments can nicely be illustrated referring to the representative mass increase measurement depicted in Fig. 6.1. Four distinct regimes are indicated. In the beginning the sample is put into the thermostatting cell and is mounted on the laboratory scale so that it hangs freely above the bulk reservoir. Regime (a) is now described by a mass decrease of the sample, which can be seen as the obvious manifestation of its thermalization. Due to the elevated temperature (here $T = 59\,°C$) the sample loses a part of its preadsorbed water and accordingly gets lighter. A sufficient thermalization is then given for an adequate convergence of the mass towards m_s.

The measurement is started by moving the cell upward using the height-adjustable cantilever until the sample touches the liquid surface. Immediately a meniscus forms along the block's perimeter. This induces a force F_S towards the reservoir acting on the sample, which becomes noticeable by a sudden jump in mass right at the beginning of regime (b). The connection between F_S and the perimeter length P is determined by

$$F_S = P \sigma \cos\theta_0 \tag{6.1}$$

widely known from the Wilhelmy plate assembly for measuring the surface ten-

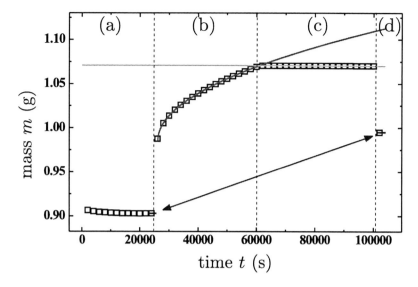

Figure 6.1.: Representative measurement of the mass increase of a porous Vy-cor® block (V10, $V_s = 6.25 \times 5.05 \times 12.70\,\text{mm}^3$) due to the imbibition of n-tetracosane (n-$\text{C}_{24}\text{H}_{50}$) at $T = 59\,°\text{C}$. The red line is a $\sqrt{t}$-fit to the liquid's imbibition behavior, the green line describes the level of saturation. The two blue lines represent the sample's mass before (m_s) and after (m_e) the experiment. The overall mass increase is then $M = m_e - m_s$. Such an experiment can very well be divided into 4 distinct regimes as indicated by the dashed lines and marked by (a) - (d). The data density is reduced by a factor of 2000.

sion of a liquid. Here it enables one to quantitatively analyze the mass jump. Assuming $\theta_0 = 0°$ and given the surface tension $\sigma \approx 30\,\frac{\text{mN}}{\text{m}}$ of n-tetracosane at $T = 59\,°\text{C}$ Eq. (6.1) gives $F_S \approx 0.7\,\text{mN}$ or, equivalently, an apparent mass jump of $\Delta m \approx 0.07\,\text{g}$. This is in high accordance with the measurement. Yet, this jump can also be less distinctive since the counteracting buoyancy force sensitively depends on the actual depth of immersion.

Anyhow, this jump is only an initial effect and can be seen as a constant offset that should not disturb the subsequent imbibition process itself. The latter is the outstanding effect in regime (b). It can be described by a $\sqrt{t}$-fit in accordance to Eq. (4.13). Thus, the $\sqrt{t}$-fit provides C_m and then Eq. (4.13) results in the imbibition strength

$$\Gamma = \frac{C_m}{\rho A}\sqrt{\frac{2\eta}{\phi_i \sigma \cos\theta_0}} \equiv \Gamma_{m(\text{ass})} \tag{6.2}$$

where the initial porosity ϕ_i according to Eq. (4.8) and the bulk fluid parameters as listed in Appendix A are applied. The contact angle is assumed to be $0°$ for all measured liquids and consequently $\cos\theta_0 = 1$.

At some point a level of saturation is entered. The deviation from the $\sqrt{t}$-law in regime (c) signals that the sample is completely filled. To extract its corresponding mass m_e the sample has to be detached from the bulk liquid reservoir.

For this purpose the cell is moved downward. After the separation the sample's bottom has to be cleaned with a tissue in order to get rid of any droplet that has remained there after the detachment. The blue line in regime (d) finally indicates the mass m_e of the completely filled sample. The overall mass uptake is then $M = m_e - m_s$.

In addition to the before-mentioned method one can apply the rise level relation Eq. (4.12) in order to gain access to the imbibition strength of the sample. Despite my previous statement that there is no well-defined quantity that one can call *the* rise level $h(t)$ the imbibition strength can be deduced from the time t_0 that it takes for the complete filling of a sample with height h_0. For this purpose it is most practical to identify the intersection of the $\sqrt{t}$-fit (red line) with the fit of the level of saturation (green line) as the finishing time t_0 of the rise dynamics. With this one arrives at:

$$\Gamma = \frac{h_0}{\sqrt{t_0}} \sqrt{\frac{2\phi_i \eta}{\sigma \cos\theta_0}} \equiv \Gamma_{\text{t(ime)}} \,. \tag{6.3}$$

Though, it is important to keep in mind that Γ_t is not derived from the rise level dynamics, but from one single known point: $h(t_0) = h_0$. The imbibition strength Γ_m according to Eq. (6.2) should therefore imply the more reliable results. Nonetheless, in some special cases Eq. (6.3) will be rather useful.

6.1.2. Neutron Radiography Measurements

Neutron radiography measurements permit a space- and time-resolved investigation of the liquid distribution within the sample (as opposed to the gravimetric measurements, which lack the spatial resolution). By this means not only information on the imbibition dynamics, but also on by-passing material transports via the vapor phase can be deduced. In order to assess the influence of vapor condensation beyond the advancing imbibition front, additional experiments with not or only partly sealed samples were performed. Some of these matrices were mounted above the liquid level in the reservoir so that they were only subjected to the liquid's vapor. This measuring principle renders possible a visualization of the influence of condensation processes on the overall dynamics.

In order to ensure a simultaneous start of the imbibition process all samples must be brought to the same level (except for the matrices just mentioned that should only be subjected to the vapor phase). Furthermore, all blocks have to be aligned with respect to the incoming neutron beam direction. This is important to guarantee identical transition paths through the samples and hence a uniform absorption rate over the whole matrix cross-section. Finally the sample cell has to be brought within the detection area of the CCD chip. This can be done employing the manipulator and it can be checked via some short-time transmission images.

The measurement is started by switching on the external precision pump. That way the initially empty sample cell is gradually filled with the liquid. It takes about 11 minutes at the highest throughput rate until the level reaches the bottom facets of the samples and the imbibition process starts. At that moment the pump is switched off again.

For each radiography picture the CCD chip is exposed for a constant time interval of 30 seconds. This time scale is a compromise between large collecting times for good statistics on the one hand and short collecting times for better dynamics resolution on the other hand.

A typical series of such pictures is shown in Fig. 6.2 for water invading V10. The images are normalized and rendered in pseudocolors in such a way that only changes in the liquid distribution with respect to the complete liquid-free sample cell are visible. This allows one to impressively visualize the liquid and vapor dynamics in the Vycor® samples. Even with the naked eye it is noticeable that imbibition apparently occurs much faster in the completely unsealed matrix in the middle. However, this sample additionally shows significant liquid condensation beyond the advancing front, which is, of course, responsible for the increased

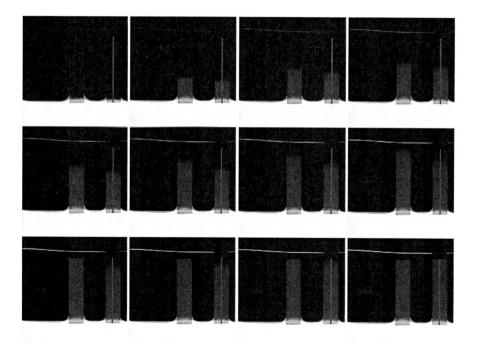

Figure 6.2.: Series of processed radiography pictures of water invading three V10 blocks. They correspond to time intervals of approximately 15 minutes. Each picture was divided by a picture from the beginning of the experiment when there is no liquid in the cell at all. The highlighting of variances is enhanced by rendering the image in pseudocolors. The right sample's top and side facets are sealed as opposed to the sample in the middle. Both blocks are 20 mm high. The left Vycor® sample (10 mm height) has only sealed side facets and hangs 10 mm above the water reservoir. The red line indicates the sample's long axis along which gray scale value profiles were taken (see Fig. 6.3).

dynamics. The importance of taking into account the condensation processes for a volatile liquid is corroborated by the left sample, which shows notable liquid uptake although it is subjected to the vapor only.

A systematic analysis of the image sequence requires the extraction of gray scale value profiles along the sample's long axis (red line in Fig. 6.2). This was done using the Java-based image processing program ImageJ. Some of the obtained profiles are shown in Fig. 6.3. As opposed to the optical measurements (see Fig. 7.3) these profiles are closely related to the actual liquid distribution in the sample.

The bump at the sample's bottom is caused by the macroscopic liquid meniscus at the sample's perimeter. Its shape is subject to time dependent variations due to the fall of the liquid level because of the finite evaporation of the bulk reservoir and the liquid transport to the sample. As a result, the bulge's occurrence at the beginning of the measurement and its gradual disappearance are mere artefacts of the normalization method with the completely filled sample and requires no further attention.

The wandering of the liquid front as well as its shape are clearly recognizable. Already with the naked eye one perceives that the front itself is not sharp but rather washed out – an effect that might be connected with the light scattering phenomenon and will be considered again in chapter 7. What is more, the profiles

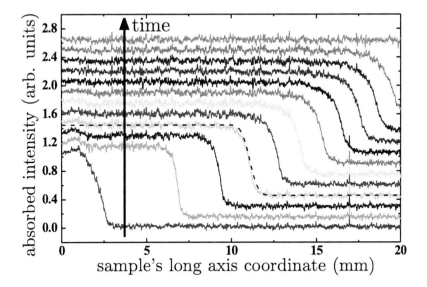

Figure 6.3.: Temporal evolution of profiles of water invading a sealed V10 sample extracted from the radiography images shown in Fig. 6.2 (right sample). The profiles are taken along the sample's long axis and are arithmetically averaged over the complete sample width. They are normalized with respect to the completely filled sample and shifted against each other by a constant value (0.15) for a better illustration of the temporal evolution. The dashed line is a fit of Eq. (6.4) to the respective profile.

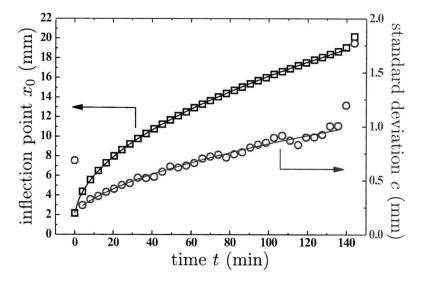

Figure 6.4.: Inflection point x_0 ($\square$) and standard deviation c ($\bigcirc$) from fits of Eq. (6.4) to the profiles partly shown in Fig. 6.3 as a function of the time t. The inflection point may be identified as the mean front position of water invading the sealed V10 sample whereas c quantifies its washout. Please note that different scalings are used. The lines are $\sqrt{t}$-fits to the underlying data. The data density is reduced by a factor of 5.

enable one to observe the liquid content beyond the advancing front. For a completely sealed sample this information reveals whether there is significant condensation caused by the evaporating front that might influence the overall dynamics comparable to the external invasion of vapor from the bulk reservoir.

In order to arrive at a more quantitative description of the behavior each profile was characterized analytically. Because of their sigmoidal shape a Gauss error function is fitted to the normalized absorbed intensity I as a function of the sample's long axis coordinate x. For this purpose the following relation was used:

$$I(x) = a - \frac{b}{2} \operatorname{erf}\left(\frac{x - x_0}{c}\right) . \tag{6.4}$$

Figure 6.3 exemplarily includes such a fit. The fitting parameters a, b, c, and x_0, as extracted from a script-based analysis using the data analyzing tool Genplot, can all be connected to a physical quantity. The presumably most important parameter in this context is the inflection point x_0. It can be identified as a mean position of the front. Its washout is comprised in the parameter c, which quantifies the standard deviation of the underlying normal distribution. Both parameters are plotted as a function of the time t in Fig. 6.4. When one compares the measured data sets with the underlying fits, it becomes evident that both quantities obey a $\sqrt{t}$-law. In consequence, the rise level h for a given time t is determined by $h(t) = x_0(t) = C_h \sqrt{t}$ resulting in

$$\Gamma = C_h \sqrt{\frac{2\phi_i \eta}{\sigma \cos\theta_0}} \equiv \Gamma_{\text{r(adiography)}} \tag{6.5}$$

based on Eq. (4.12). Again, the contact angle $\theta_0 = 0°$ and the bulk fluid parameters as listed in Appendix A are applied in order to calculate Γ_r. Unfortunately, this method provides no opportunity for an easy determination of the initial porosity comparable to Eq. (4.8) for the mass increase experiments. Thus, I will apply ϕ_i as obtained from the gravimetric measurements with adapted error bars, which account for this uncertainty.

The parameters a and b comprise the levels of the completely empty $\left[I(x \to \infty) = a - \frac{b}{2} \equiv I_\infty\right]$ and full $\left[I(x \to 0) = a + \frac{b}{2} \equiv I_0\right]$ sample parts, respectively. This permits one to examine liquid condensation processes beyond the front. Such an effect would result in $I_\infty > 0$ whereas it should remain zero if the neutron transmission rate is not modified with respect to the empty sample.

The systematic deviations of both x_0 and c for the highest t shown in Fig. 6.4 illustrate a general difficulty of the above-mentioned evaluation method. After the upper bound of the front reaches the top facet of the sample the fits of the subsequent profiles are rather poor because of the lack of the I_∞-level. Therefore this method always requires a cut-off of the results. Moreover, the deviations of c for low t toward higher front widths can conclusively be interpreted. These variations are an artefact of the measuring method. Because of the finite collecting time for each picture (30 s) the front (even if ideally sharp) becomes blurred if the front speed is high enough to traverse far more than one pixel of the CCD chip ($\sim 16\,\mu m$). Since the front gradually decreases its speed during the measurement this effect only comes into play at the very beginning of the experiment. For that reason it does not influence the $\sqrt{t}$ behavior shown in Fig. 6.4.

6.2. Water in Porous Vycor®

Water transport in nano- and mesoscale environments plays a crucial role for phenomena ranging from clay swelling, frost heave, and oil recovery to colloidal stability, protein folding, and transportation in cells and tissues [73–78]. The mobility of water has extensively been studied by experiment and theory in such restricted geometries over the last three decades.

These studies revealed a remarkable fluidity of water down to nanometer and even subnanometer spatial confinement [5–7] and demonstrated the validity of macroscopic capillarity concepts at the nano- and mesoscale [9, 10]. However, how capillary forces along with the retained fluidity of water and other liquids contribute to the huge variety of phenomena, for which self-propelled fluid transport is encountered in nano- and mesoscale geometries, has so far almost solely been investigated in theory or for simple geometries like thin films [34, 57, 79–81].

6.2.1. Gravimetric Measurements

Experiments on the capillary rise of water in V5 and V10 were conducted at three different temperatures T (25 °C, 40 °C, and 60 °C). The results are summarized in Tab. 6.1. The imbibition strengths were deduced according to Eq. (6.2) from the mass increase dynamics (Γ_m) and according to Eq. (6.3) from the time t_0 that it takes for the complete filling of the sample (Γ_t). As already stated earlier, for the characterization of the imbibition dynamics Γ_m should provide the more reliable value since it is directly derived from the dynamic behavior, namely $m(t)$. In contrast Γ_t is based on one single point only.

Anyhow, the results for both V5 and V10 reveal a remarkable variance of Γ_m whereas Γ_t is – as predicted for the imbibition strength – nearly independent of the temperature T. The values of Γ_m are always larger than the respective ones of Γ_t and the absolute deviations tend to increase with the temperature. Yet, this peculiar behavior is an artefact of the measuring method, which can conclusively be explained referring to a fundamental flaw in the gravimetric measurements for highly volatile liquids. The liquid level of the bulk reservoir gradually declines due to evaporation and consequently the depth of immersion of the sample decreases. This results in a steady reduction of the buoyancy force, which is equivalent to an additional apparent mass increase: the mass uptake dynamics and accordingly Γ_m are seemingly enhanced. Of course, t_0 and hence also Γ_t are not influenced by this effect and as the evaporation rate increases with the temperature the excess of Γ_m with respect to Γ_t is more distinctive at elevated temperatures.

In summary it can be said that in this special case the imbibition dynamics are systematically erroneously described by Γ_m. Due to this fact I have to refer to Γ_t instead. According to Tab. 6.1 the imbibition strengths are then given by $(123.7 \pm 2.1) \cdot 10^{-7} \sqrt{\mathrm{m}}$ for V5 and $(173.7 \pm 3.1) \cdot 10^{-7} \sqrt{\mathrm{m}}$ for V10 arithmetically averaged over all temperatures.

Table 6.1.: Characterization of the imbibition dynamics of water in porous Vycor® at three different temperatures by means of imbibition strengths as deduced from gravimetric measurements. Both Γ_m according to Eq. (6.2) and Γ_t according to Eq. (6.3) are listed (in units of $10^{-7} \sqrt{\mathrm{m}}$).

temperature	V5		V10	
	Γ_m	Γ_t	Γ_m	Γ_t
25 °C	186.2 ± 8.5	126.7 ± 8.2	227.3 ± 10.7	170.0 ± 7.9
40 °C	148.8 ± 7.8	120.5 ± 7.4	253.9 ± 12.9	180.9 ± 11.6
60 °C	257.2 ± 11.5	125.3 ± 11.2	273.5 ± 13.6	174.3 ± 11.4

Table 6.2.: Characterization of the imbibition dynamics of water in porous Vycor® at room temperature by means of the imbibition strength Γ_r as deduced from neutron radiography measurements according to Eq. (6.5) (in units of $10^{-7} \sqrt{\mathrm{m}}$).

	V5	V10
sealed	120.0 ± 5.9	166.3 ± 7.2
unsealed	156.6 ± 8.9	199.6 ± 10.6

6.2.2. Neutron Radiography Measurements

The imbibition strengths Γ_r of water invading V5 and V10 for an unsealed as well as for a completely sealed sample are listed in Tab. 6.2. Here, the initial porosities ϕ_i as obtained from the gravimetric measurements (0.29 ± 0.01 for V5 and 0.30 ± 0.01 for V10) were applied. A comparison with the just stated Γ_t values reveals a suitable coincidence for the completely sealed samples.

The results again reflect the before-mentioned enhanced imbibition dynamics of the unsealed samples. The excess amounts to a 20% to 30% increase, which is closely related to a material influx via the vapor phase beyond the advancing front. This is illustrated by the profiles' levels I_0 and I_∞ shown in Fig. 6.5. For the sealed samples I_∞ remains zero during the whole measurement meaning that the neutron transmission is not modified at all. As a consequence, no vapor condenses beyond the front. This is contrary to the unsealed samples, which are characterized by a gradual increase in I_∞ because of a by-passing material transport via the vapor phase.

The systematic deviations of I_∞ for the highest t again reflect the general deficiency of the evaluation method after the upper bound of the front reaches the

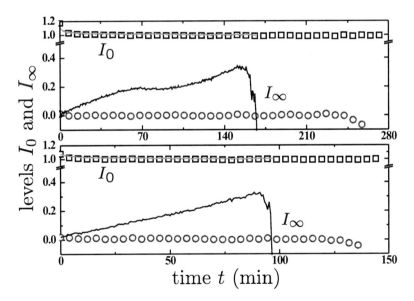

Figure 6.5.: Levels I_0 ($\square$ and —) and I_∞ ($\bigcirc$ and —) of density profiles of water invading V5 (upper panel) and V10 (lower panel) as obtained from neutron radiography measurements. Squares and circles correspond to the completely sealed sample, solid lines indicate the unsealed sample. Values $I_0 > 1$ for $t \to 0$ and their subsequent convergence towards unity are the manifestation of the already before-mentioned initial occurrence of a bulge in the profiles, which hampers the fitting of the I_0-level. The data density is reduced by a factor of 5 (V10) and 10 (V5), respectively.

top facet. This flaw has already been pointed out in the preceding discussion on the temporal behavior of c and x_0 (see Fig. 6.4). Moreover, the peculiar behavior of I_∞ for V5 (as compared with I_∞ for V10) is an artefact of a non-standard sample preparation. The clamping of this sample required the sealing of a strip of the side facets. This sealing in turn prevented the material influx via the vapor phase at a certain level, which is clearly recognizable as a temporary stagnation of I_∞.

6.2.3. Discussion

The imbibition strengths $\Gamma = (122.8 \pm 1.7) \cdot 10^{-7} \sqrt{m}$ for V5 and $\Gamma = (172.4 \pm 2.8) \cdot 10^{-7} \sqrt{m}$ for V10 were consistently ascertained observing two different manifestations of the same process: both mass increase and rise level measurements provided comparable results. For one thing, this enhances the plausibility of the deduced quantities and the likewise deduced discoveries. For another thing, the nanoscopic model of the imbibition dynamics as introduced in chapter 4 is confirmed. This is a rather important issue concerning the reliability of all subsequent results.

A basic assumption was proven to be true: there is no by-passing material transport via the vapor phase, provided appropriate measures are taken. The impact of unimpeded vapor influx is imposingly illustrated by the enhanced dynamics shown in Fig. 6.2 and quantified in Tab. 6.2. Nonetheless, evaporation from the advancing front cannot be prevented by a sealing. Based on thermodynamic argumentations its influence was estimated to be merely small (see section 4.3). This assumption is verified by the results shown in Fig. 6.5: during a capillary rise experiment there is no change of the water content beyond the advancing front at all. Thus, the measured dynamics can solely be attributed to the capillarity-driven liquid invasion even for highly volatile liquids such as water. What is more, as predicted from the theory of the capillary rise dynamics, the imbibition strength was proven to be no function of the temperature.

Reconsidering Eq. (4.14) one is able to calculate the hydrodynamic pore radius r_h from the measured imbibition strengths Γ. To this end I assume $r_L = (r_0 - 2.5\,\text{Å}) \pm 2.5\,\text{Å}$ as stated in section 4.2 and the matrix properties of V5 and V10 according to Tab. 3.1. With this in mind one arrives at $r_h = (2.85 \pm 0.21)$ nm for V5 and $r_h = (4.41 \pm 0.28)$ nm for V10, or, in terms of the slip length $b = r_h - r_0$,

$$b \approx -5\,\text{Å} \quad \text{for water invading V5 and}$$
$$b \approx -5\,\text{Å} \quad \text{for water invading V10.}$$

Consistently for both V5 and V10 a *negative* slip length of about 5 Å was found. Considering the diameter of a water molecule of approximately 2.5 Å this result can be interpreted as follows: two layers of water directly adjacent to the pore walls are immobile meaning that they are pinned and do not take part in the flow. Taking this into account (through the above-mentioned negative slip length) the imbibition dynamics can conclusively be described by the model proposed in chapter 4. In consequence one may conclude that the residual inner compartment of the liquid obeys classical hydrodynamics based on continuum mechanical theory. As a result, my experiments confirm former findings on the conserved fluidity [5–7] and capillarity [9, 10] of confined water – except for the pinned layers.

The assumption of the compartmentation of mesopore-confined water just stated is corroborated by recent molecular dynamics studies on the glassy structure of water boundary layers in Vycor® and the expected existence of sticky boundary layers in Hagen-Poiseuille nanochannel flows for strong fluid-wall interactions [82–87]. By means of X-ray diffraction distortions of the hydrogen-bonded network of water near silica surfaces were found [88], which might be responsible for the markedly altered liquid properties. Tip-surface measurements document a sudden increase in the viscosity by orders of magnitude in 0.5 nm proximity to hydrophilic glass surfaces [89]. It also extends former experimental results with respect to the validity of the no-slip boundary condition for water/silica interfaces, in which this condition was proven down to at least 10 nm from the surface [90] whereas slip-flow for water is only expected at hydrophobic surfaces [90, 91].

Here, it is important to highlight that a significant influence of silica dissolution in water can definitely be ruled out. The resultant aqueous electrolyte (often referred to as polywater) would of course be characterized by altered liquid properties and could therefore likewise be responsible for the observed decreased overall invasion dynamics. Nevertheless, as was pointed out in detail by Tombari *et al.* [92], the dissolution of silica in water for the measuring times and temperatures considered here is vanishingly small and may be neglected.

The presented imbibition experiments entail additional rheologic details. As mentioned earlier Eqs. (4.12) and (4.13) intrinsically assume a parabolic velocity profile across the pore cross section, which implies a linear variation in the viscous shear rate $\dot\gamma$, starting with $0\,\frac{1}{s}$ in the pore center to a t-dependent maximum at r_h: $\dot\gamma_{max} \propto \frac{1}{\sqrt{t}}$. For the V5-experiment, for example, one may estimate $\dot\gamma_{max}$ to decrease from $7\cdot10^4\,\frac{1}{s}$ after 1 s to $3\cdot10^2\,\frac{1}{s}$ at the end of the experiment. Since there are no t-dependent, and consequently no $\dot\gamma$-dependent, deviations of $m(t)$ from a single $\sqrt{t}$-fit the measurements testify to the absence of any non-Newtonian behavior of water. Despite the relatively large $\dot\gamma$'s ascertained here, the latter is not too surprising. The viscous forces of $\mathcal{O}(\eta\,d^2\,\dot\gamma)$ (with the molecular dimension d on the order of a few Å) can only overcome the strong water/silica interactions of $\mathcal{O}(A/d)$ (with the Hamaker constant $A \sim 10^{-19}$ J) for $\dot\gamma > 10^{12}\,\frac{1}{s}$ [30, 93] – significantly beyond the shear rates probed here.

Finally, it is worthwhile noting that water encounters a negative pressure, which linearly decreases from $-p_L$ (on the order of some hundred bar) at the advancing menisci to atmospheric pressure at the bulk reservoir. Water's hydrogen bridge bond network is expected to be responsible for an increase in viscosity η and decrease in density ρ under such large tensile pressures. Based on thermodynamic models for stretched water [94, 95], one may estimate a T-dependent $\sim 3\%$ decrease in the overall dynamics due to this effect, which is, unfortunately, below the error margins.

6.3. Hydrocarbons in Porous Vycor®

The imbibition study of water in porous Vycor® glass just presented reveals many aspects of the behavior of a liquid in interface-determined systems. Anyhow, these results could just as well have been obtained on a planar glass substrate, meaning that so far no effects explicitly due to the liquids' restrictions to a cylindrical pore geometry have been reported. One might wonder whether the water molecules are too small as compared with the channel diameter to be subject to any limitations. This consideration motivates the hereinafter presented systematic study of the imbibition dynamics of chain-like hydrocarbons namely the homologous series of n-alkanes (n-C_nH_{2n+2}) and the branched hydrocarbon squalane.

The normal alkanes are one of the most basic organic series. They are building blocks for surfactants, liquid crystals and lipids. The alkane molecule's all-trans length ℓ is proportional to the number of C atoms n in the chain's backbone

$$\ell(n) = 2.1\,\text{Å} + (n-1) \cdot 1.25\,\text{Å} \tag{6.6}$$

whereas their diameter is independent of n (approximately $3\,\text{Å}$). Hence, the alkane series allows one to easily vary the aspect ratio of the molecule, while many important liquid properties (density ρ, surface tension σ, viscosity η) nearly remain the same (see Tab. A.4 for detailed values). In particular all alkanes totally wet silica. For that reason, alkanes are particularly well suitable for a systematic study of the influence of the shape of the liquids' building blocks on the flow dynamics through channels with diameters comparable to the molecule's length.

6.3.1. Gravimetric Measurements

Linear hydrocarbons ranging from n-decane (n-$C_{10}H_{22}$) to n-hexacontane (n-$C_{60}H_{122}$) being tantamount to all-trans molecule lengths ℓ between 1 nm and approximately 8 nm were applied. Furthermore, the branched hydrocarbon squalane

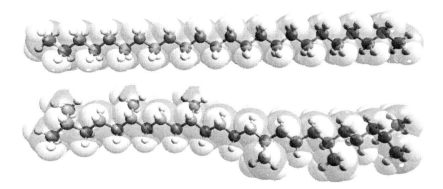

Figure 6.6.: Illustrations of the all-trans configurations of the linear hydrocarbon n-tetracosane (n-$C_{24}H_{50}$) and the branched hydrocarbon squalane ($C_{30}H_{62}$). The backbones of both chains carry 24 C atoms, so they coincide in the all-trans molecule length $\ell \approx 3$ nm.

(2,6,10,15,19,23-hexamethyltetracosane) was used. The six additional methyl side groups as compared to n-tetracosane (see Fig. 6.6) significantly influence the liquid properties. This fact can impressively be exemplified referring to their melting points (squalane: $-40\,°C$, n-tetracosane: $50\,°C$). However, the altered structure and thickness of the squalane molecule might affect the flow dynamics in extreme spatial confinement as well.

Gravimetric measurements were performed with eleven hydrocarbons in both V5 and V10 at the temperatures stated in Tab. A.4. For better statistics the imbibition dynamics of each liquid/substrate combination were ascertained at least three times. Contrary to the water measurements no remarkable variance in Γ_m could be detected. Considering the 15 times higher vapor pressure of water ($p_0 = 23.4\,\text{mbar}$) compared to n-decane ($p_0 = 1.6\,\text{mbar}$) at $20\,°C$ this is not too surprising: the evaporation of the bulk reservoir is not sufficient to induce an apparent enhancement of the rise dynamics due to a decrease of the buoyancy force.

The results of the study are summarized in Fig. 6.7. It is clearly recognizable that the measured imbibition strengths can be described with the arithmetic averages over all measured n-alkanes $\overline{\Gamma}_m = (124.8 \pm 2.8) \cdot 10^{-7}\,\sqrt{\text{m}}$ for V5 and

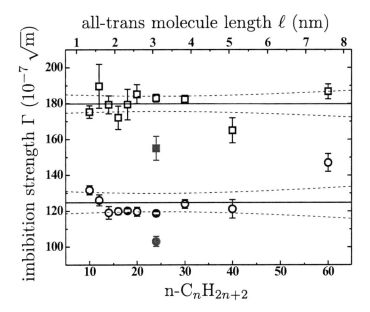

Figure 6.7.: Imbibition strengths Γ_m obtained from gravimetric measurements for a series of linear hydrocarbons (open symbols) ranging from n-decane up to n-hexacontane invading V10 ($\square$) and V5 ($\bigcirc$), respectively. Their mean values are visualized by the solid lines, lying within the confidence interval (dashed lines, confidence level: 90 %). The corresponding all-trans molecule length ℓ of the alkane can be read from the top axis. The measured branched hydrocarbon squalane is indicated by the filled symbols.

Table 6.3.: Imbibition strengths Γ_m of n-hexadecane invading V5 as a function of the relative humidity (RH). In addition the initial porosities ϕ_i as extracted from the mass increase measurements are listed.

RH (%)	ϕ_i	Γ_m $(10^{-7}\sqrt{m})$
24	0.275	116.5 ± 3.8
34	0.262	121.9 ± 5.3
50	0.245	121.9 ± 4.1

$\overline{\Gamma}_m = (179.8 \pm 2.3) \cdot 10^{-7}\sqrt{m}$ for V10, respectively. For a wide range of molecule lengths ℓ there is no influence on the imbibition dynamics at all. This statement is corroborated by the inserted confidence intervals that broaden only slightly and merely for the highest ns, thus they suggest that the 'true' Γ is obtained in an ℓ-invariant interval. Significant deviations can only be observed for n-hexacontane (n-$C_{60}H_{122}$) in V5.

An additional interesting result are the decreased dynamics of squalane. The imbibition strength is lowered for both V10 (14 % lessening) and V5 (18 % lessening). In comparison with the n-alkane of equal length (n-tetracosane) the six extra side groups seem to significantly influence the liquid's dynamic behavior in pore confinement.

Besides, the invasion of n-hexadecane (n-$C_{16}H_{34}$) in V5 was gravimetrically (and optically) recorded as a function of the relative humidity (RH) of the surrounding atmosphere. For this purpose the whole setup as depicted in Fig. 5.1 (along with the CCD camera) was put into a box. Applying a humidifier the water content of the atmosphere within this housing could be increased to a maximum RH of approximately 50 %. The resultant imbibition strengths Γ_m are listed in Tab. 6.3 together with the extracted values of the initial porosities. No distinct influence of the humidity on the imbibition dynamics is detectable, provided one takes into account the gradual and significant decrease in ϕ_i with increasing humidity. The results of the optical measurements will be addressed in chapter 7.

6.3.2. Neutron Radiography Measurements

The imbibition dynamics of n-tetradecane (n-$C_{14}H_{30}$) invading V5 and V10 at room temperature were additionally recorded by means of neutron radiography experiments. Based on the initial porosities ϕ_i as obtained from the gravimetric measurements (0.275 ± 0.01 for V5 and 0.295 ± 0.01 for V10) one arrives at $\Gamma_r = (118.6 \pm 4.0) \cdot 10^{-7}\sqrt{m}$ for V5 and $\Gamma_r = (172.8 \pm 5.6) \cdot 10^{-7}\sqrt{m}$ for V10. Again, these values coincide within the error margins with the mean values of the gravimetric measurements just stated.

Moreover, the measurements verify that for the observation of the invasion dynamics of hydrocarbons a sealing of the sample's side facets is not required at all. This is illustrated by the profiles' levels I_0 and I_∞ shown in Fig. 6.8. Even

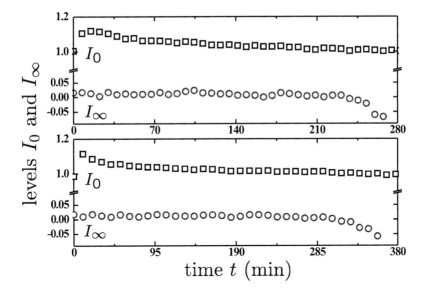

Figure 6.8.: Levels I_0 ($\square$) and I_∞ ($\bigcirc$) of density profiles of n-tetradecane invading unsealed samples of V5 ($h_0 = 10\,\text{mm}$, upper panel) and V10 ($h_0 = 15\,\text{mm}$, lower panel) as obtained from neutron radiography measurements. Values $I_0 > 1$ for $t \to 0$ and their subsequent convergence towards unity are the manifestation of the already before-mentioned initial occurrence of a bulge in the profiles, which hampers the fitting of the I_0-level. The data density is reduced by a factor of 15 (V10) and 10 (V5), respectively.

though the samples were not sealed I_∞ remains zero during the whole measurement. According to that, no vapor condenses beyond the front.

6.3.3. Discussion

Both mass increase and rise level measurements of unbranched hydrocarbons in porous Vycor® show comparable results. The invasion dynamics can be expressed through the imbibition strengths $\Gamma = (122.8 \pm 2.9) \cdot 10^{-7}\,\sqrt{\text{m}}$ for V5 and $\Gamma = (178.8 \pm 2.5) \cdot 10^{-7}\,\sqrt{\text{m}}$ for V10. Here, no distinct influence of the all-trans chain length ℓ on the respective n-alkane's invasion dynamics can be detected. Only for the largest molecule n-hexacontane ($\ell \approx 7.6\,\text{nm}$) in the smallest pores of V5 ($r_0 = 3.4\,\text{nm}$) hints for significant deviations from the arithmetic average over all measured hydrocarbons towards increased overall imbibition dynamics are found $\left[\Gamma = (147.1 \pm 5.0) \cdot 10^{-7}\,\sqrt{\text{m}}\right]$. Interestingly, enhanced dynamics for long-chain molecules (mostly polymers) have already been documented in the literature [22, 34, 96] and could be ascribed to substantial slip at the liquid-substrate interface. This possibility will be addressed later on.

Again, the hydrodynamic pore radii r_h can be deduced from the measured imbibition strengths Γ. One attains $r_h = (2.85 \pm 0.22)\,\text{nm}$ for V5 and $r_h = (4.49 \pm$

0.28) nm for V10, or in terms of the slip length $b = r_h - r_0$

$$b \approx -5\,\text{Å} \quad \text{for n-alkanes invading V5 and}$$
$$b \approx -4\,\text{Å} \quad \text{for n-alkanes invading V10.}$$

Comparable to the water measurements for both V5 and V10 a *negative* slip length of approximately $5\,\text{Å}$ is found.

In this context it is worthwhile considering the impact of the pore walls' initial water coating on the liquid invasion. According to the results presented in Tab. 6.3 there is no significant influence of the RH on the overall rise dynamics, provided one takes into account the initial porosity as a function of the RH. However, especially the first adsorbed water layer is highly stabilized by the attractive potential between silica and water [46]. In addition to that, a water coating on the pore walls is favored compared to an alkane coating. This is expressed by their different surface tensions and can be demonstrated by displacement measurements. Here, an initially full hydrocarbon filling in porous Vycor® is replaced by a water filling through imbibition (after connecting the sample to a water reservoir). Cross checks of the sample's mass change with the different liquid densities reveal an absolutely complete substitution of water for the alkane.

One has to assume that the imbibed hydrocarbon is not able to entirely displace the initial water coverage. At least the first layer adjacent to the pore walls should be present during the capillary rise process. This extremely immobile layer provides a first contribution of $\sim 0.25\,\text{nm}$ to the overall sticking layer thickness stated above. But there is still a residual of b, which can be ascribed to either a second water layer or to a layer of flat lying hydrocarbon molecules. Thanks to their similar thicknesses one cannot distinguish between these two cases beforehand.

Yet, the concept of a layer of flat lying hydrocarbons is confirmed by the squalane measurements. The $\sim 15\,\%$ decrease in the imbibition strength and the consequent additional reduction of the hydrodynamic pore radii $[r_h = (2.61 \pm 0.21)\,\text{nm}$ for V5 and $r_h = (4.19 \pm 0.32)\,\text{nm}$ for V10$]$ by $\sim 0.3\,\text{nm}$ can uniformly be explained by the larger diameter of the squalane molecule as compared with a non-branched hydrocarbon chain. The extra methyl groups attached to the n-tetracosane backbone cause an increase of the sticking layer's thickness by a value comparable to the decrease in r_h just mentioned. Interestingly, a similar influence of the additional side branches on the boundary conditions was already reported [16]. The continuously positive slip lengths b for both n-hexadecane and squalane revealed within this FRAP study notwithstanding, b was found to be always lower for the branched squalane.

The sticky hydrocarbon layer assumption is further corroborated by the imbibition dynamics being influenced by neither the relative humidity in the laboratory (see Tab. 6.3) nor the temperature of up to $110\,°\text{C}$ during the n-$C_{60}H_{122}$ measurement (see Fig. 6.7). Both quantities directly affect the degree of the initial water coverage down to the strongly bound first monolayer of water in the high-T experiments. Anyhow, the dynamics are not influenced at all. This result is especially important regarding Ref. [97], in which already the presence of one to two water monolayers preadsorbed on a silica surface was found to be sufficient to considerably reduce the London force field of the glass. Nevertheless, the measured invasion dynamics prove that the modified alkane-substrate interactions do not change the wetting behavior of the hydrocarbon.

Summarizing, one may assume the sticking layer to be constituted of a water layer directly adjacent to the pore walls and a contiguous layer of flat lying hydrocarbon chains whereas the innermost liquid compartment obeys classical hydrodynamics based on continuum mechanical theory. Not until the chain length of the hydrocarbon reaches the diameter of the pore the results are seemingly influenced by the respective shape of the liquids' building blocks.

The assumption of a sticky monolayer of hydrocarbons is confirmed by forced imbibition experiments on n-alkanes in Vycor® [4], by studies regarding the thinning of n-alkane films in the surface force apparatus [5, 98–100] and by molecular dynamics simulations [82, 101]. On top of this, X-ray reflectivity studies indicate one strongly adsorbed, flat lying monolayer of hydrocarbons on silica [102–104]. The retained viscosity of the confined alkanes was also verified by means of MD simulations [8].

With that the just stated results disagree with a series of recent publications. Especially the existence of a *positive* slip length b of short-chain alkanes was proven several times by experiment [12, 16] as well as by simulation [8, 105], even for complete wetting and hence highly attractive liquid-substrate interaction. Interestingly, the hydrodynamic pore radius $r_h = (3.12 \pm 0.26)\,\text{nm}$ of n-hexacontane in V5 reveals a slip length $b = -0.28\,\text{nm}$, which is, admittedly, still negative. Nonetheless, it is markedly increased as compared with $b = -0.55\,\text{nm}$ for the arithmetic average over all n-alkanes by approximately $3\,\text{Å}$ – the thickness of the layer of flat lying hydrocarbons. This result might be interpreted as a step towards a slip boundary condition. Yet, because of the relatively large error bars it must be treated as a mere indication of altered boundary conditions.

6.4. Silicon Oils in Porous Vycor®

So far the results on the flow of water and linear hydrocarbons in extreme spatial confinement revealed a retained fluidity and capillarity of the liquids. Considering the highly attractive interaction between fluid and substrate the verification of a sticking layer is not too surprising. Liquids slipping at a substrate are mainly observed in non-wetting configurations [11, 17, 90, 91]. Unfortunately such systems cannot be investigated by means of spontaneous imbibition, which always requires a wetting liquid (the forced throughput measurements in Part III are not subject to such restrictions).

However, applying polymeric or oily liquids also seems to facilitate the occurrence of slip at the fluid-solid interface [22]. This presumption is corroborated by a recent imbibition study of polystyrene [96]. What is more, the MD simulations carried out in Ref. [34] demonstrate a slip boundary condition for the capillary rise of a short-chain macromolecule (representative of the building blocks of silicon

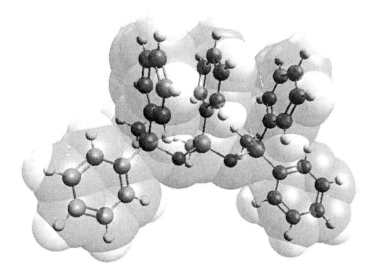

Figure 6.9.: Illustration of the molecular structure of 1,1,3,5,5-Pentaphenyl-1,3,5-trimethyltrisiloxane. It is the main constituent of the applied Dow Corning diffusion pump oils DC704 and DC705. Five phenyl and three methyl groups are regularly attached to the trisiloxane ($Si_3O_2-R_8$) backbone. The molecular dimensions can roughly be estimated using the diameter of a phenyl ring ($\sim 5.4\,\text{Å}$) to be between $1\,\text{nm}$ and $1.5\,\text{nm}$.

oil) with slip lengths on the order of $30\,\%$ of the tube radius. Therefore, when one tries to observe modified boundary conditions it will be rather promising to use such liquids.

For this purpose I applied the diffusion pump oils DC704 and DC705 from Dow Corning. Their main constituent is the siloxane depicted in Fig. 6.9 with molecular dimensions between $1\,\text{nm}$ to $1.5\,\text{nm}$. They are characterized by a negligible vapor pressure and a relatively high viscosity (see Tab. A.3). Because of the latter the capillary rise process is substantially slowed down. The BCLW-$\sqrt{t}$ behavior can be observed for weeks. Again, this preeminently elucidates the retained fluidity and Newtonian character of the liquid. The extracted imbibition strengths Γ_{m} are listed in Tab. 6.4.

Compared to the values of water or the hydrocarbons the dynamics of the silicon oils are decreased. This is equivalently expressed in terms of the resultant slip lengths b (see Tab. 6.4). They all indicate a sticking layer boundary condition whose thickness is approximately $1.4\,\text{nm}$. Interestingly this result is independent of both the sample type and the silicon oil. Thus it is not too erroneous to conclude that, again, one layer of siloxane molecules is pinned to the pore walls and does thereby substantially lower the invasion dynamics.

This result does not necessarily have to contradict the findings of the studies mentioned at the beginning [22, 34, 96]. The applied silicon oils presumably are not chain-like enough but rather block-like. Possibly this internal molecular

Table 6.4.: Characterization of the imbibition dynamics of Dow Corning silicon oils in porous Vycor® at room temperature by means of the imbibition strength Γ_m (in $10^{-7}\sqrt{m}$) and the resultant slip length b (in nm).

DC...	V5		V10	
	Γ_m	b	Γ_m	b
704	63.6 ± 5.6	-1.35 ± 0.23	109.2 ± 12.6	-1.39 ± 0.40
705	53.2 ± 4.8	-1.52 ± 0.21	122.4 ± 11.4	-1.19 ± 0.38

structure is decisive for the occurrence of slip at the interface. In this regard one has to recall the increased dynamics of n-hexacontane in V5 compared to the short-chain hydrocarbons. This result stimulates future studies on more extended alkane and polymer melts with molecular weights beyond $1\,\frac{kg}{mol}$.

6.5. Conclusion

The results of the systematic study of the capillary rise of water, hydrocarbons and silicon oils in a network of silica mesopores just presented unambiguously reveal a compartmentation of the pore confined liquid. With high accuracy one perceives an interfacial boundary layer adjacent to the pore walls with a defined thickness whose dynamics are mainly determined by the interaction between liquid and substrate. This manifests itself in terms of negative slip lengths.

Independent of the respective liquid a monolayer of water directly adjacent to the pore walls is most likely an essential part of this sticking layer. It is present right from the start due to the finite humidity in the laboratory. Besides, it is highly stabilized and cannot be displaced by any of the liquids investigated. The remaining part of the sticking layer thickness can be attributed to a second pinned layer composed of molecules of the respective invading liquid. Here, the chain-like hydrocarbons are arranged parally to the pore walls (flat-lying).

For the residual inner region (away from the interface) classical concepts of continuum mechanical theory and, consequently, flow dynamics predictions based on collective liquid properties such as viscosity and surface tension remain valid. Furthermore, for the liquids investigated and in the shear rate region explored no significant rate dependence of the viscosity could be detected. The interesting impact of the high tensile pressures on the viscosity and the density of water could not be assessed since they entail only little deviations, which are well below the error margins.

Additionally, I found hints for the influence of the shape of the liquids' building blocks on the mesoscopic flow behavior. For one thing the slowed down dynamics of a branched hydrocarbon compared with an unbranched one is a mere consequence of the molecules enlarged thickness and the consequent enlargement of the pinned boundary layer [85]. What is even more surprising are the enhanced invasion dynamics for the longest hydrocarbon chain investigated in the smallest pores applied within this study. This result might be interpreted as a first hint of the importance of molecular slip during spontaneous imbibition.

Finally, I was able to disprove any alternative invasion mechanism that can compete with the capillary rise process. Especially the material influx via the vapor phase could be excluded, provided appropriate measures are taken. The predictions of the merely small influences of evaporation processes from the advancing front based on thermodynamic argumentations in section 4.2 were impressively verified. Therefore the measured dynamics can solely be attributed to the capillarity-driven liquid invasion even for highly volatile liquids such as water.

7. INVASION KINETICS IN POROUS VYCOR®

The roughening of fronts between wet and dry regions in disordered media is a common phenomenon. The spatial and temporal statistics of the interfaces are easily observed by the naked eye, as shown in Fig. 7.1 or in Fig. 7.2, and they remind, in general terms, of other 'rough' objects such as fractures in inhomogeneous media, fronts in slow combustion, flux lines in disordered superconductors, and Brownian paths in diffusion [75]. In this chapter parts of the measurements of the capillary rise in porous Vycor® previously presented will be investigated with regard to the dynamics of the interface.

7.1. Interface Roughening during Spontaneous Imbibition

The dynamics of interfaces and fronts in disordered systems are affected by noise. Part of it stems from ever-present thermal fluctuations, while the rest is due to the 'randomness' of the porous medium, that is quenched disorder. For porous Vycor® glass such 'randomness' is provided through the pore size distribution. This so-called capillary disorder acts at the interface in terms of local variations of the Laplace pressure and thereby causes a blurring of the invasion front.

Figure 7.1 illustrates this behavior in the presence of a wide radii distribution (right panels). For narrow distributions the interfaces remain sharp. Moreover, the invasion speed v, or equivalently the capillary number $\mathrm{Ca} \propto v$, seemingly plays a decisive role as well: roughening only occurs at capillary numbers well below 10^{-4}. For higher invasion speeds the interface again remains more or less sharp.

7.1.1. Scaling Law of the Vertical Front Width

Interestingly, many measurable quantities with regard to the blurring of interfaces obey simple scaling laws [107]; so does the vertical width w of the interface, which increases as a power of time t

$$w(t) \propto t^{\beta} . \tag{7.1}$$

The exponent β is called growth exponent and characterizes the time dependent dynamics of the roughening process. It is this front width $w(t)$, and hence β that can easily be deduced from my imbibition measurements: one must only observe the advancing imbibition front as a function of time. According to a recent phase field model study of the interface roughening in spontaneous imbibition the growth exponent was found to be $\beta \approx \frac{1}{3}$ in the limit of low Ca [108]. As outlined in subsection 4.6.2 this limit holds for my experiments as well.

7.1.2. Lateral Correlations in Interface Roughening during Spontaneous Imbibition

It is important to note that the $\left(\beta = \frac{1}{3}\right)$-scaling of the front width as predicted from the phase field model in Ref. [108] purely originates in the physics of the imbibition process itself. Fluid is transported towards the front from the reservoir behind. Therefore, advanced parts of the interface receive less, whereas retarded parts receive more liquid than the average. Furthermore, since the pores are interconnected they partly compete for liquid coming from the same region behind the front, thus the collective advancement speed of such regions is restricted. This

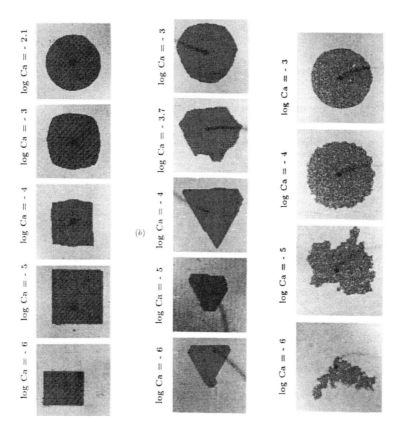

Figure 7.1.: Stable imbibition of injected oil (black areas) displacing air (bright areas) in various pore networks as a function of the capillary number Ca. (left panels): Square network with a narrow distribution of channel width. (middle panels): Triangular network with a narrow distribution of channel width. (right panels): Square network with a wide distribution of channel width (from 0.16 mm to 0.64 mm). Taken from Ref. [106] by courtesy of Roland Lenormand, Institut Francais du Petrole, Rueil-Malmaison, France.

demand is a direct consequence of the law of conservation of mass, which must be fulfilled.

Eventually, the competition for liquid results in a lateral correlation in the roughening process: due to the network effects the pores cannot fill independendly. The limitation of available liquid and the natural deceleration of far advanced parts as well as the acceleration of retarded parts of the interface consequently lead to a stabilization of the imbibition front. To put it in a nutshell, thanks to the interconnectivity of the pores the front should smooth itself and the blurring should be limited. Over the course of this chapter I will show that a potential lack of such correlations implicates exponents larger than $\beta = \frac{1}{3}$ and correspondingly much more blurred interfaces.

7.2. Experimental Results

By measuring the width of the roughened interface in the conducted imbibition experiments the scaling law just presented can be assessed. Principally two methods can be applied. Optical measurements gain access to the roughened interface via the induced light scattering imposed by the refractive index contrast between empty and filled pore spaces as soon as they have length scales characteristic of the wavelengths of the employed light. Nevertheless, multiple scattering processes render the interpretation of these measurements particularly difficult. Therefore, the neutron radiography measurements, which measure the neutron absorption and hence are directly sensitive to the local liquid concentration, yield results that are more straightforward to interpret. Nonetheless, comparing the optical results with gravimetric and neutron radiography experiments will possibly unravel the microscopic structures that are responsible for the light scattering. For that reason both methods were applied.

7.2.1. Optical Measurements

A typical series of pictures taken during a standard imbibition experiment is shown in Fig. 7.2. The white front is clearly recognizable. In order to gain information on its dynamics a systematic analysis of the pictures has to be done. To this end gray scale value profiles along the sample's long axis (red line in Fig. 7.2) were elaborated. This was done using the Java-based image processing program ImageJ. Based on the profiles shown in Fig. 7.3 the dynamics can easily be extracted.

As already mentioned the front's width gradually increases during the experiment. Thus, a complete description has to comprise two aspects of the dynamics: the front's position and its width or, equivalently, the front's upper and lower bound both as a function of the time. The latter information can be gathered from the profiles, in which the front appears as a broad peak wandering from left to right. Its boundaries can be ascertained by a simple comparison of the respective profile with reference profiles of the complete empty and full sample, respectively. In doing so, the front manifests itself by unambiguous deviations from the reference, which can systematically be detected employing a script-based analysis using the data analyzing tool Genplot.

The results from this analysis method are shown in Fig. 7.4. The included fits suggest that both the upper and the lower boundary of the light scattering front advance $\propto \sqrt{t}$. For convenience the observed behavior can again be characterized by means of a imbibition strength Γ. However, one has to keep in mind that these values reveal nothing about the microscopic structure of the matrix as opposed to the gravimetric and the neutron radiography measurements presented in the last chapter. They solely serve a better comparability with other measurements.

In accordance to Eq. (4.12) and with the $\sqrt{t}$-law's proportionality constants C_{ub} and C_{lb} of the front's upper and lower bound, respectively, the above-mentioned measurements will be characterized via

$$\Gamma = C_{ub/lb} \sqrt{\frac{2\,\phi_i\,\eta}{\sigma\,\cos\theta_0}} \equiv \Gamma_{ub/lb}\,. \tag{7.2}$$

The values extracted from optical measurements with water and n-tetradecane in V5 as well as in V10 at room temperature are listed in Tab. 7.1. Additionally, for n-hexadecane in V5 such measurements were systematically carried out as a function of the relative humidity (RH) in the laboratory. For this purpose a humidifier was applied, which permits an artificial increase of the relative humidity.

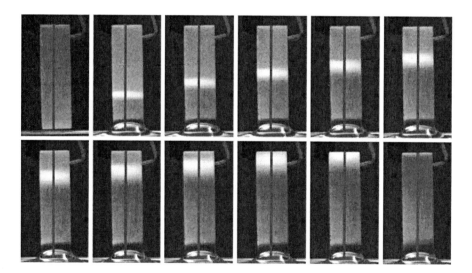

Figure 7.2.: Series of pictures of n-tetradecane (n-$C_{14}H_{30}$) invading a porous Vycor® block (V10, $V_s = 4.55 \times 4.55 \times 14.90\,\text{mm}^3$) at room temperature. The first picture shows the sample immediately before the start of the experiment. Each of the following photographs reveals the situation one hour later until the sample is completely filled (in the last picture) after approximately 620 minutes. The red line indicates the sample's long axis along which gray scale value profiles were taken (see Fig. 7.3).

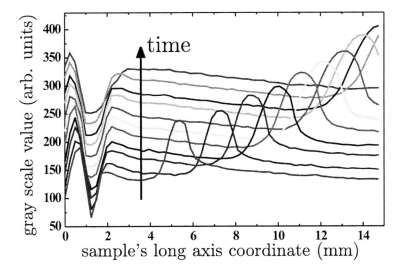

Figure 7.3.: Temporal evolution of gray scale value profiles of the pictures shown in Fig. 7.2. The profiles are taken along the sample's long axis and are arithmetically averaged over the complete sample width. They are shifted against each other by a constant value (20) for a better illustration of the temporal evolution. The white front is clearly recognizable as the broad peak wandering from left to right. The dip on the left side corresponds to the macroscopic meniscus along the block's perimeter.

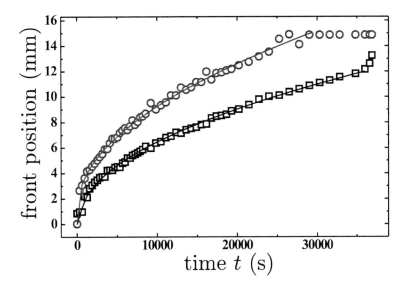

Figure 7.4.: Time dependent upper (○) and lower (□) bound position of the front extracted from the profiles partly shown in Fig. 7.3. The lines are $\sqrt{t}$-fits to the underlying data. The data density is reduced by a factor of 20.

Table 7.1.: Imbibition strengths Γ_{ub} of the upper bound and Γ_{lb} of the lower bound (in $10^{-7}\sqrt{m}$) of the advancing light scattering front as extracted from optical rise level measurements of water, n-tetradecane, and n-hexadecane invading V5 and V10 at $T = 25\,°C$. The hexadecane measurements were performed as a function of the relative humidity (RH, in %) in the laboratory.

sample	fluid	RH	Γ_{ub}	Γ_{lb}
V5	H_2O		119.1 ± 5.6	95.0 ± 5.1
	n-$C_{14}H_{30}$		117.5 ± 6.4	94.8 ± 5.4
	n-$C_{16}H_{34}$	24	122.7 ± 5.8	99.2 ± 5.3
		34	122.4 ± 6.2	90.5 ± 5.6
		50	119.9 ± 7.6	84.3 ± 6.2
V10	H_2O		170.3 ± 7.4	136.4 ± 6.6
	n-$C_{14}H_{30}$		178.8 ± 8.2	139.1 ± 7.2

Standard values of RH in the laboratory range from 20 % to 30 %. They were recorded by means of an electronic hygrometer with internal data logger.

7.2.2. Neutron Radiography Measurements

The measuring and evaluation methods applied for the neutron radiography experiments have already been explained in detail in subsection 6.1.2. Over the course of this evaluation I introduced the parameter c, which can be taken as a measure of the washout of the advancing front (to be exact, c is equal to $\sqrt{2}$ times the standard deviation of the underlying normal distribution around x_0). In Fig. 6.4 this parameter is plotted as a function of the time t. In comparison to the underlying fit it is obvious that c also obeys a $\sqrt{t}$ law.

For a more convenient relation to the values Γ_{ub} and Γ_{lb} obtained from the optical measurements one can again calculate an imbibition strength from the $c(t)$ behavior. With the proportionality coefficient C_c of the $\sqrt{t}$-fit one arrives at

$$\Gamma = C_c\sqrt{\frac{2\,\phi_i\,\eta}{\sigma\cos\theta_0}} \equiv \Gamma_c\,. \tag{7.3}$$

An upper bound of the front is then given by $\Gamma_r + \Gamma_c$ and the lower bound by $\Gamma_r - \Gamma_c$ with the values of the imbibition strength Γ_r according to the results from the previous chapter. In Tab. 7.2 the respective values from the conducted experiments with water and n-tetradecane in V5 and V10 are summarized.

7.3. Discussion

The discussion of the results is split into two parts. The first part considers the scaling behavior of the interface. The growth exponent β is ascertained and

Table 7.2.: Imbibition strengths $\Gamma_r + \Gamma_c$ of the upper bound and $\Gamma_r - \Gamma_c$ of the lower bound (in $10^{-7}\sqrt{m}$) of the washed-out front as extracted from neutron radiography measurements of water and n-tetradecane invading V5 and V10 at $T = 25\,°C$.

sample	fluid	$\Gamma_r + \Gamma_c$	$\Gamma_r - \Gamma_c$
V5	H_2O	124.5 ± 5.9	115.5 ± 5.9
	n-$C_{14}H_{30}$	123.2 ± 4.0	114.0 ± 4.0
V10	H_2O	172.9 ± 7.2	159.7 ± 7.2
	n-$C_{14}H_{30}$	183.1 ± 5.6	162.5 ± 5.6

interpreted within the framework of the theoretical conceptions introduced in the beginning of this chapter. In contrast, the second part addresses the dynamic behavior of the interface roughening, meaning the prefactor of the scaling law Eq. (7.1).

7.3.1. Discussion: Part I – Scaling Behavior

In accordance with Fig. 7.4 and Fig. 6.4 the advancing imbibition front was found to roughen with the square root of time – $w(t) \propto t^{\frac{1}{2}}$ – by both measuring methods and for all liquids applied in both sample types. This renders the result rather unambiguous, though it is in disagreement with the prediction of the phase field model in Ref. [108], which suggests $\beta = \frac{1}{3}$.

A first scientific approach for an explanation of the observed $\sqrt{t}$ behavior can be gained from the so-called random deposition (RD) model [107] as illustrated in Fig. 7.5 (left). It is the simplest imaginable growth model and it suggests that a particle falls vertically from a randomly chosen site over the surface until it reaches the top of the column under it, whereupon it is deposited. The characteristic feature of this model is that the roughness is completely uncorrelated: the columns grow independently since there is no mechanism that can generate correlations along the interface – the correlation length is always zero.

Such lateral correlations are considered in the ballistic deposition (BD) model as shown in Fig. 7.5 (right), e.g., employing the illustrated nearest neighbor sticking rule. Interestingly, the BD model yields $\beta = \frac{1}{3}$ in accordance with the phase field model [108] whereas the RD model yields $\beta = \frac{1}{2}$ in accordance with my imbibition study. The far more pronounced interface roughening in the RD model is illustrated in the lower panel of Fig. 7.5.

This discovery for a network of multiply connected pores is rather bewildering. It reflects, however, results from gas permeation measurements on systematically modified porous Vycor® and porous silicon samples (see section B.4). Here, the observed diffusion coefficients suggest that, to a considerable extent, Vycor® glass behaves analogous to an array of independent pores. In addition, the subsequent

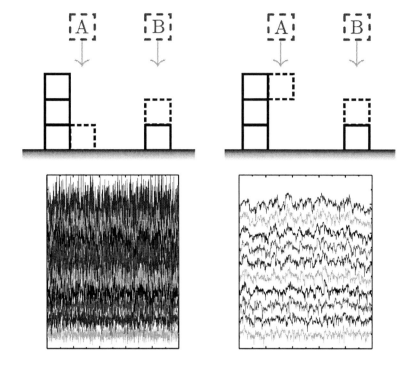

Figure 7.5.: (upper panel): Illustration of the random deposition (left) and the ballistic deposition model (right, with nearest neighbor sticking rule). (left): Particles A and B are dropped from random positions above the surface and are deposited on the top of the column under them. (right): Again A and B are dropped from random positions above the surface but now the particle sticks to the first site along its trajectory that has an occupied nearest neighbor. (lower panel): Temporal evolution of interfaces generated on the basis of the random deposition (left) and the ballistic deposition model (right). It consists of 1,000 columns. Between each interface line 10,000 particles are additionally deposited.

discussion on the dynamic behavior of the front roughening gives further results that can suitably be interpreted referring to the filling of independent pores rather than a multiply connected pore network.

7.3.2. Discussion: Part II – Dynamic Behavior

In this second part of the discussion I will turn to the dynamics of the front roughening, i.e. the absolute degree of the front width. This is easily possible by means of the imbibition strengths previously calculated. A comparison between the values stated in Tab. 7.1 and in Tab. 7.2 reveals some distinct differences in the obtained signals. These will be considered in the following.

Some first useful results can be drawn from the comparison of the values of the upper bound. For this purpose Γ_{ub} and $\Gamma_r + \Gamma_c$ are recollected in Tab. 7.3. From this contrasting juxtaposition it is evident that the advancement speed of the upper bound of the light scattering front is only slightly, but systematically smaller than the upper bound as extracted from neutron radiography experiments. Reconsidering the results from the gravimetric study one may even conclude that the light scattering front's upper bound nearly coincides with the mass increase front. This means that only below the 'mean' imbibition interface there are structures of percolating clusters that generate light scattering. It is therefore by no means a signal of vapor condensing beyond the front; but this has already been excluded so far.

For more details I will turn to the absolute width w of the front. For the optical measurements it is given by $\Gamma_{ub} - \Gamma_{lb}$, for the neutron radiography experiments it is equal to $2\Gamma_c$. Both values are summarized in Tab. 7.4; for n-hexadecane also as a function of the relative humidity (RH) in the laboratory.

Here the basic differences in the observed signals become evident. The front widths as extracted from the light scattering signal exhibit almost twice the values obtained from neutron scattering. Furthermore, a systematic behavior is reflected in both signals: the width in V5 is always smaller than in V10. The

Table 7.3.: Summary of the results of the dynamics of the upper bounds of the imbibition front as extracted from optical (Γ_{ub}) and neutron radiography ($\Gamma_r + \Gamma_c$) measurements. All values in units of $10^{-7}\sqrt{m}$.

sample	fluid	Γ_{ub}	$\Gamma_r + \Gamma_c$
V5	H_2O	119.1	124.5
	n-$C_{14}H_{30}$	117.5	123.2
V10	H_2O	170.3	172.9
	n-$C_{14}H_{30}$	178.8	183.1

Table 7.4.: Summary of the results of the dynamics of the width w of the imbibition front as extracted from optical ($\Gamma_{ub} - \Gamma_{lb}$) and neutron radiography ($2\Gamma_c$) measurements. All values in units of $10^{-7}\sqrt{m}$.

sample	fluid	RH	$\Gamma_{ub} - \Gamma_{lb}$	$2\Gamma_c$
V5	H_2O		24.1	9.0
	n-$C_{14}H_{30}$		22.7	9.2
	n-$C_{16}H_{34}$	24	23.5	
		34	31.9	
		50	35.6	
V10	H_2O		33.9	13.2
	n-$C_{14}H_{30}$		39.1	20.6

influence of the liquid itself is not as clearly assessable: for V5 it seems to be independent of the liquid, for V10 the width for water is smaller than that for tetradecane. This inconsistent behavior might be a fact attributable to another quantity that seems to significantly influence the front width: with increasing RH in the laboratory the width increases as well. But, it is important to note that the RH does influence neither the speed of the upper bound (see Tab. 7.1) nor the overall imbibition dynamics (see Tab. 6.3). Unfortunately the values of RH were not recorded as standard and for that reason it cannot be assessed afterwards to what extent humidity can explain this ambiguous behavior for different liquids.

What are the reasons for the significantly larger extent of the light scattering front? I can give an answer to this question referring to a more sophisticated evaluation method of the radiography data. I will make use of its sensitivity to the local liquid concentration. For this purpose detailed knowledge of the exact shape of the advancing interface is required. The standard fit of a symmetric profile according to Eq. (6.4) can certainly give some helpful clues on the advancement speed and the washout of the interface. But they are only exact if the front's shape actually is symmetric. Considering Fig. 7.6 it is obvious that this applies not at all. The following evaluation method does not contain such flaws.

Thanks to normalization the gray scale value profiles shown in the inset of Fig. 7.6 (or Fig. 6.3) directly coincide with the local filling degree $0 \leq f \leq 1$ of the sample

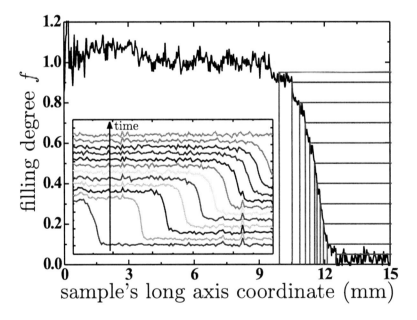

Figure 7.6.: Illustration of the evaluation method of the normalized gray scale value profiles obtained from neutron radiography measurements of n-tetradecane invading V10. The shown grid corresponds to the following filling degrees f: 5 %, 10 %, 20 %, ..., 80 %, 90 %, 95 %. Inset: Series of normalized gray scale value profiles of n-tetradecane invading porous Vycor® (identically equal to Fig. 6.3).

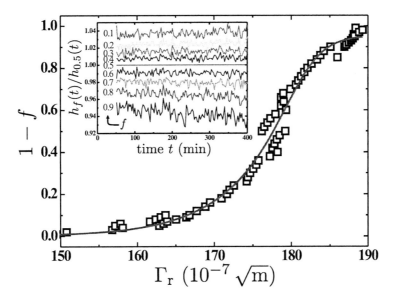

Figure 7.7.: Inverse filling degree $1 - f$ as a function of the imbibition strength Γ_r ($\square$) for n-tetradecane invading V10. The solid line is a polynomial fit to the underlying data set, which is used for an analytical investigation of the $f(\Gamma_r)$-relation. Inset: Selected rise level curves $h_f(t)$ for given filling degrees f normalized by $h_{f=0.5}(t)$. The respective filling degrees are listed within the plot.

at a certain level h. One must note that the attenuation of the incoming neutron beam is essentially exponential (similar to the Beer-Lambert law in optics). Nonetheless, for the small sample thicknesses of the Vycor® blocks and the liquids regarded here the intensity can be considered to decay linearly.

The washed-out interface between the already completely filled ($f = 1$) and the still empty part ($f = 0$) of the sample is then analyzed employing the grid method illustrated in Fig. 7.6. For a whole series of filling degrees between zero and unity I extracted the coordinate h_f for which the respective level f has been traversed by the profile for the first time. Repeating this procedure for all profiles eventually results in a series of rise level curves $h_f(t)$.

It is not too surprising that all these rise level curves again obey a $\sqrt{t}$ law. This fact is verified by the $h_f(t)$-curves in the inset of Fig. 7.7. Getting rid of the $\sqrt{t}$ dependency by normalization with the respective measurement for $f = 0.5$ it is evident that they all follow the same dynamic behavior. Hence, applying Eq. (6.5) one can again calculate imbibition strengths Γ_r from the C_h-values obtained from $\sqrt{t}$-fits to the rise level curves $h_f(t)$. Consequently, for each filling degree f one ends up with a corresponding imbibition strength $\Gamma_r(f)$. The systematic implementation of this evaluation method finally provides detailed information on the imbibition dynamics parametrized by a close relation between Γ_r and f. Such a relationship is shown in Fig. 7.7.

The chosen representation in terms of $1 - f$ instead of f simplifies the subsequent evaluation. It can easily be seen that $1 - f$ is the integrated probability of finding a pore with any Γ_r of maximum the corresponding imbibition strength $\Gamma_r(1 - f)$. In consequence, one can deduce a probability distribution function of imbibition strengths by simply calculating the first derivative of $1 - f$ with respect to Γ_r. For this purpose a polynomial fit to the $(1 - f)$-Γ_r-relationship was applied. The resultant distributions $P(\Gamma_r)$ are shown in Fig. 7.8 for n-tetradecane invading both V5 and V10 (solid lines).

The distribution functions preeminently illustrate the asymmetric shape of the advancing front. Indeed, $P(\Gamma_r)$ reveals a remarkable resemblance with the pore radii distributions in Fig. 3.2 (see Fig. 7.8 for a direct comparison). They collectively own the shape being composed of a most probable value with a more steep decrease up to larger values and a rather sustained decay down to smaller values, thereby constituting a distinct asymmetry.

Furthermore, the close relation between the pore radius r and the imbibition strength Γ via Eq. (4.14) allows for a more fundamental analysis of the resemblance between their distribution functions. With the transformations $r_h \rightarrow r + b$

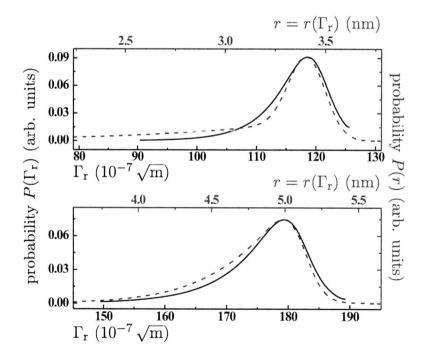

Figure 7.8.: Probability distribution functions $P(\Gamma_r)$ (solid lines) of the imbibition strengths Γ_r for n-tetradecane invading V5 (upper panel) and V10 (lower panel), respectively. The pore size distributions $P(r)$ (dashed lines, identically equal to Fig. 3.2) are also plotted in the figure. The r-scaling (top axis) is connected to the Γ_r-scaling (bottom axis) via Eq. (7.4) with $b = -6\,\text{Å}$ (V5) and $b = -4.3\,\text{Å}$ (V10), respectively. The curves were normalized to the peak value.

(see subsection 2.2.2) and $r_L \rightarrow r - 2.5\,\text{Å}$ (see section 4.2) one obtains

$$\Gamma(r) = \frac{(r+b)^2}{r} \sqrt{\frac{\phi_0}{\tau\,(r - 2.5\,\text{Å})}} \quad \propto \sqrt{r}\,. \tag{7.4}$$

Along with the membrane properties ϕ_0 and τ according to Tab. 3.1 and the slip lengths b as stated in the previous chapter this relation enables one to *quantitatively* compare the probabilities for a certain Γ and the respective radius $r = r(\Gamma)$. The distribution functions of Γ_r and r, or equivalently the bottom and top axes, shown in Fig. 7.8 are connected through this relation.

Interestingly, there is not only a qualitative resemblance between both distributions, but also a quantitative one – or, explained in an intelligible manner: each existent pore radius r (meaning each pore radius with a probability different from zero) implicates a corresponding measurable imbibition strength Γ. Vice versa, the probability for any randomly chosen Γ is directly proportional to the probability of the corresponding radius $r(\Gamma)$.

Anyhow, since slowly filling smaller pores should be bypassed via rapidly filling larger ones one would expect that the Γ distribution eventually narrows and possibly becomes more symmetric – to put it in a nutshell, the interface should smooth and no Γ corresponding to neither the smallest nor the largest existing pores should be found. In principle, this argumentation coincides with the basic assumptions of the scaling prediction outlined in subsection 7.1.2.

However, such a narrowing does obviously not happen and therefore, again, my findings imply that the porous Vycor® glass behaves as if it was made up of many independent pores: one can easily reproduce the shape of the liquid concentration profiles in Fig. 7.6 assuming a set of aligned and non-interconnected capillaries obeying the pore size distributions in Fig. 7.8. What is more, many discoveries on the roughening dynamics mentioned earlier can conclusively be explained within the framework of this model.

The systematically broader interface for V10 as compared with V5 is a direct consequence of such behavior. Since $\Gamma \propto \sqrt{r}$ (see Eq. (7.4)) it is evident that because of the larger pore radii in V10 a certain distribution width Δr in r generates broader variations in Γ for V10 than for V5.

The observation of the dynamics of the light scattering interface is a method being sensitive on structures composed of percolating paths already filled with liquid whereas others are still empty. Thinking in terms of rapidly filling large and slowly filling small pores the emergence of this phenomenon is nothing more than a logical consequence. Comparing both Γ_{ub} and Γ_{lb} of the n-tetradecane front in V5 and V10 (see Tab. 7.1) with the f-Γ_r-relationships as obtained from the radiography measurements (see Fig. 7.7) one can directly connect the emergence of the light scattering phenomenon to a corresponding filling degree f of the sample.

For both V5 and V10 the upper bound of the light scattering front Γ_{ub} is located at $f \approx 0.45 \pm 0.05$, meaning that for lower filling degrees too less pores are filled creating characteristic variation length scales of the refractive index higher than the wavelength of visible light. Crossing the $f \approx 0.45$ bound structures on adequate length scales are gradually created.

The 'center of mass' (COM) of the interface must be located in the proximity of $f \approx 0.5$, for which already filled parts and still empty ones balance each other. With the preceding discovery on the position of the upper light scattering bound in mind one may conclude that the latter can be equated to the COM location. It is in turn this COM location that must be obtained from the previously presented gravimetric experiments, which measure an integrated signal over the whole sample (and for that reason lack the spatial resolution as mentioned earlier). Indeed, as was pointed out in the very beginning of this part of the discussion, such a consistency was found, thus my interpretation is confirmed.

A remarkable fact is that appropriate light scattering structures are seemingly present up to rather high filling degrees as indicated by the lower bound's imbibition strengths Γ_{lb}. Actually, according to the radiography data the sample should already be completely filled – but a completely filled matrix does not scatter any visible light.

This flaw can consistently be traced back to a fundamental inaccuracy of the radiography results because of the salient signal noise partly caused by gamma radiation from the reactor core (gamma spots). Accordingly, one has to reckon with a decrease in resolution for filling degrees in the proximity of the flat plateaus of the I_0- and the I_∞-level being tantamount to $f \to 1$ and $f \to 0$ (see Fig. 7.6).

Indeed, the optical measurements reveal that there are still empty pores within the samples building appropriate light scattering structures down to rather low Γs. I cannot exclude that there are even smaller imbibition strengths since the disappearance of the light scattering can, again, be explained with too large variation length scales of the few still empty pores. With regard to the Γ_r distributions in Fig. 7.8 this new insight implicates that the slow decay down to smaller Γs is presumably even more sustained, thus its resemblance with the radii distributions is reinforced.

Finally, as already mentioned earlier, it is obvious that even relative humidities of up to 50 % do not distinctly influence the dynamics of the mass increase or (as was just pointed out) equivalently the upper bound of the light scattering front, provided one takes into account the gradual and significant decrease in the initial porosity ϕ_i with increasing humidity (see Tab. 6.3).

Interestingly, the lower bound of the light scattering front noticeably slows down with rising humidity. Within my model this is not too surprising since the impact of a coverage with a certain thickness affects smaller pores far more than larger pores. Whereas this effect does not seem to influence the many larger pores at all it has a distinct impact on the small pores, which show decreased dynamics and, thereby, extend the light scattering front to lower values. But, since the small pores are only few and carry only little fluid at all this phenomenon does not influence the overall dynamics of the rise process.

7.4. Conclusion

The presented detailed investigation on the interface dynamics suggests the absence of any network effect during the imbibition process. The scaling as well as the dynamic behavior of the driven front indicate the lack of any lateral correlation at the advancing interface that might smooth the roughness. One may conclude that the pore connectivity plays a subordinate role only. This discovery reflects results from gas permeation measurements on systematically modified porous Vycor®, which are presented in section B.4. Here, the observed diffusion coefficients suggest that, to a considerable extent, porous Vycor® glass behaves analogous to an array of independent pores (porous silicon).

The reasons for this surprising finding are a matter of conjecture. One might only speculate whether the small aspect ratios of the pore sections, i.e. the relatively large pore length as compared with their diameter, along with the monodirectional liquid flow, i.e. the main flow direction pointing from the sample's bottom facet towards its top facet, prevent any lateral correlation (that is perpendicular to the main flow direction). Interestingly, network effects are very well noticeable when there is no such preferential direction in the system that 'breaks' the isotropy of the matrix. In particular this is the case for the steady state of an adsorbent, e.g., capillary condensation in Vycor® [54–56]. As elucidated in section 4.5 desorption occurs via percolating paths which entails significant light scattering: the sample turns immediately white thereby providing direct evidence of their network-like structure. Simulations of the imbibition process in a complex pore network are a promising method in order to eventually shed light on this behavior. Future studies will address this approach.

Many discoveries, e.g., broader interfaces for V10 as compared with V5, are found to be a direct consequence of the observed absence of network effects. Thinking in terms of rapidly filling large and slowly filling small pores the emergence of clusters that induce light scattering is nothing more than a logical impact. Light scattering occurs for filling degrees $f \gtrsim 0.45 \pm 0.05$ (coincident with the 'center of mass' location as deduced from the integrated signal obtained in gravimetric measurements) up to almost 1.

The extent of the white front down to extremely low Γs implicates the existence of small pores, which fill rather sedately. However, since they carry only little fluid they do not influence the overall rise dynamics at all. An increase of the RH has only consequences for this lower bound of the light scattering front. Here the impact of the additionally adsorbed water layers is the most distinctive and becomes noticeable in a further increase of the front width down to even smaller Γs. The overall rise dynamics are not influenced by the RH.

8. Phase Transitions in Porous Vycor®

In this final chapter on spontaneous imbibition the influence of the mesopore confinement on the behavior of some temperature induced phase transitions of the invading liquid will be examined. Usually phase transitions are accompanied by unique variations of the fluid properties. Since the overall rise dynamics sensitively depend on these quantities one can easily detect such characteristic deviations. According to Eq. (4.13) the desired information is comprised in the prefactor C_m of the observed $\sqrt{t}$ behavior of the measured mass increase:

$$C_{\mathrm{m}} = \rho A \sqrt{\frac{\phi_{\mathrm{i}}\, \sigma\, \cos\theta_0}{2\,\eta}}\, \Gamma \,, \tag{8.1}$$

with only the quantities highlighted in red depending on the temperature. Here, the slight T dependence of ϕ_{i} can be neglected in the limited T regions observed. Moreover, the static contact angle θ_0 is zero for all T. The matrix properties are T-invariant and can be eliminated through normalization. Hence, by performing several imbibition measurements at different temperatures in the T-range of interest a simple comparison of the prefactors might already provide an insight into the phase transition behavior of the confined liquid.

8.1. Surface Freezing in n-Tetracosane

At first I present a T-dependent study of the capillary rise of the linear hydrocarbon n-tetracosane (n-$C_{24}H_{50}$) in porous Vycor®. The chosen alkane exhibits a surface freezing transition at $T_{\mathrm{s}} = 54\,^\circ\mathrm{C}$ which is accompanied by a characteristic reversal of the $\sigma(T)$-slope. This distinctive behavior must affect the rise dynamics, which permits one to examine the phase transition in mesopore confinement.

8.1.1. Surface Freezing Transition

Surface freezing is the formation of a single solid monolayer floating on the bulk melt in a T-range between the bulk freezing temperature T_{f} and a temperature T_{s} [109–113] (see Fig. 8.1). Upon surface freezing the hydrocarbon molecules are rectified, parallel-aligned with their long axis along the surface normal, and the center of mass lattice is hexagonal, resulting in a 3 nm thick surfactant-like monolayer. The abrupt onset of surface freezing at T_{s} allows one to switch on (and off) the 2D crystalline phase at the free surface by a small T variation. The goal

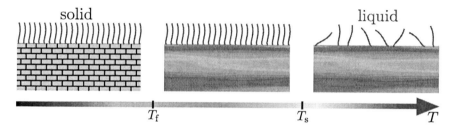

Figure 8.1.: Illustration of the surface freezing transition. Between the bulk melting point T_f and the phase transition temperature T_s one single solid mono-layer is established on the free surface of the liquid.

of this study is to investigate whether this particular molecular rearrangement is detectable by and how it affects the imbibition dynamics of the alkane in mesopores. This phase transition is known from a series of n-alkanes ranging from n-$C_{14}H_{30}$ to n-$C_{50}H_{102}$, however, I selected n-tetracosane for this study, since it exhibits the largest surface freezing T-range ($\sim 3\,°C$) of all pure n-alkanes (see Fig. 8.2).

The surface freezing transition is accompanied by a distinctive signature in the T behavior of the surface tension σ, known from bulk n-alkanes, that is a change from a small negative, above T_s, to a large positive T-slope, below T_s (see Fig. 8.3). It can be understood from elementary surface thermodynamics [109]. The surface

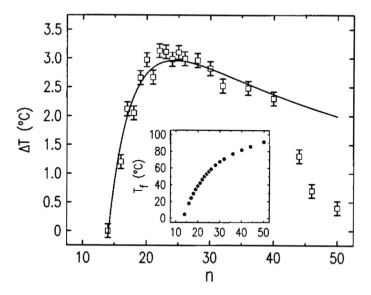

Figure 8.2.: The temperature range of existence $\Delta T = T_s - T_f$ for the surface crystalline phase as a function of the chain length in terms of the number of carbon atoms n in the hydrocarbon backbone. The solid line is a theoretical expression derived in Ref. [109], and the inset charts the bulk freezing temperatures T_f vs n. Courtesy of Moshe Deutsch, Bar-Ilan University, Ramat-Gan, Israel.

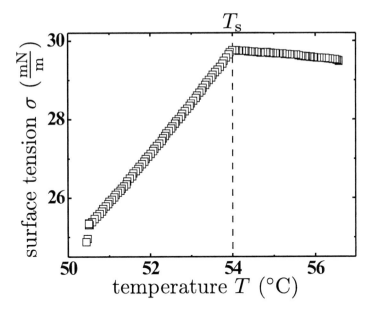

Figure 8.3.: Surface tension σ of bulk n-tetracosane as a function of the temperature T as deduced from Wilhelmy plate measurements. Courtesy of Moshe Deutsch, Bar-Ilan University, Ramat-Gan, Israel.

tension σ is a direct measure of the surface excess free energy:

$$\sigma(T) = \epsilon_s - \epsilon_b - T \cdot (S_s - S_b) \qquad (8.2)$$

where ϵ_s and ϵ_b are the energies and S_s and S_b the entropies for the surface and the bulk phase, respectively. The temperature slope of surface tension shows information on the surface excess entropy: $d\sigma/dT = -(S_s - S_b)$. The negative slope of bulk tetracosane for $T > T_s$ is typical of an ordinary liquid surface, for which the molecules on the surfaces are less constrained than those in the bulk, thus S_s is slightly larger than S_b, yielding $d\sigma/dT < 0$. Surface freezing and its first-order character result in an abrupt reduction of the surface entropy S_s in such a way that S_s is smaller than S_b leading to $d\sigma/dT > 0$.

8.1.2. Results

This distinctive behavior of the surface tension must be mapped in the T-dependence of the measured mass uptake. In Fig. 8.4 some representative measurements of tetracosane invading V5 and V10 are shown along with their corresponding $\sqrt{t}$-fits. It is clearly recognizable that the overall dynamics decrease with decreasing temperature what is mostly attributable to the increasing viscosity. Anyhow, the absolute diminishment seems not to be a linear function of the temperature T at all. The difference between 54 °C and 61 °C is obviously smaller than the difference between 52 °C and 54 °C, although the change in T is three times larger. In the following this qualitative discovery will quantitatively be analyzed referring to the prefactors obtained from the fits.

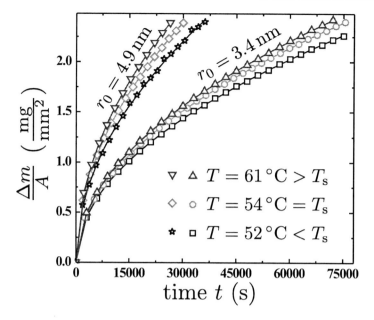

Figure 8.4.: Specific mass uptake of both V5 and V10 due to n-tetracosane imbibition as a function of the time for selected temperatures close to, but above the bulk freezing temperature. Solid lines correspond to $\sqrt{t}$-fits. The data density is reduced by a factor of 1000.

For convenience the prefactors are normalized by the prefactor value at an arbitrarily chosen temperature T^n. This procedure finally yields the normalized imbibition speed v^n of the measurement at the temperature T. One should not be confused by the term 'speed', since v^n is actually dimensionless. Nevertheless, it somehow quantifies the mass uptake rate of the sample at a given temperature. According to Eq. (8.1) this quantity can also be calculated:

$$v^n(T) \equiv \frac{C_m(T)}{C_m(T^n)} = \frac{\rho(T)}{\rho(T^n)} \cdot \sqrt{\frac{\sigma(T) \cdot \eta(T^n)}{\sigma(T^n) \cdot \eta(T)}} \, . \tag{8.3}$$

It is important to note that in Eq. (8.3) the geometry as well as the internal topology of the substrate do not play a role any longer. Assuming the $\sigma(T)$ kink-behavior according to Fig. 8.3 along with the T dependency of η and ρ (corresponding to Tab. A.4) one can calculate theoretical values of $v^n(T)$. They are plotted as a solid line in the upper panel of Fig. 8.5. The measured speeds are indicated by the single points.

8.1.3. Discussion

The quantitative agreement between measured and calculated values is reasonable. In particular for $T < T_s$ the measured imbibition speeds do not follow the dashed line, which is based on surface tension values extrapolated from the ordinary values beyond T_s (see lower panel in Fig. 8.5) and, hence, correspond to a suppression of the surface frozen state. Due to the remarkable coincidence of the temperature characteristic of the kink in v^n and the degree of deviation from a suppressed phase transition I feel encouraged to solely attribute the distinct change in the imbibition dynamics to a change in $\sigma(T)$ at the advancing menisci typical of surface freezing. The corresponding nanoscopic flow configuration is illustrated in the insets of Fig. 8.5 for both $T < T_s$ and $T > T_s$. Note that independent of the temperature the sticking layer boundary condition at the pore walls has to be applied as stated in chapter 6.

Furthermore, the slope difference in $\sigma(T)$ inferred from linear fits of the measured

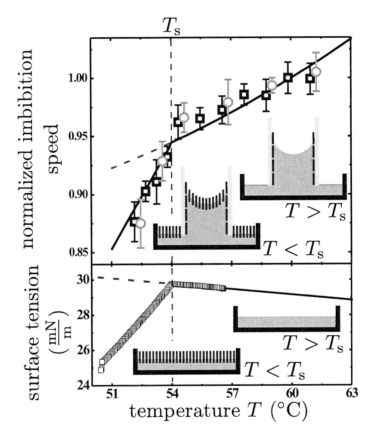

Figure 8.5.: (upper panel): Measured normalized imbibition speeds v^n (for $T^n = 60\,^\circ\mathrm{C}$) of tetracosane in V5 ($\square$) and V10 ($\circ$), respectively, in comparison to values calculated under the assumption of meniscus freezing (—). Insets: Illustration of a surface frozen layer at an advancing meniscus ($T < T_s$) and of a sticky, flat lying boundary layer at the silica pore wall in the entire T-range investigated. (lower panel): T-dependent surface tension of tetracosane by way of comparison (see Fig. 8.3 for a detailed view). Insets: Illustrations of surface freezing at a planar tetracosane surface. The dashed lines correspond to extrapolations of the calculated imbibition speeds and of the surface tension below T_s in the absence of surface freezing, respectively.

imbibition speeds below and above T_s of $1.75 \pm 0.3 \, \mathrm{mN/(m \, K)}$ is, within error margins, equal to $1.40 \pm 0.1 \, \mathrm{mN/(m \, K)}$ for surface freezing at bulk tetracosane melts. Since this quantity measures the loss of entropy of surface molecules upon entering the surface frozen state: $\Delta(\mathrm{d}\sigma/\mathrm{d}T) = \mathrm{d}\sigma/\mathrm{d}T(T < T_s) - \mathrm{d}\sigma/\mathrm{d}T(T > T_s) = \Delta S$, the measurements indicate that the molecular alignment at the advancing imbibition front leads to a loss in entropy comparable to the one observed at quiescent bulk surfaces and thus, presumably, to a well ordered 2D crystalline state. Having in mind the dynamics of the frozen meniscus along the tortuous pore path, the frequent encountering of pore junctions with irregular channel geometries and the small curvature radii of the menisci, for which large defect densities of the 2D crystals are expected [114, 115], this may seem surprising. Note, however, that the defect formation rate is presumably proportional to the monotonically in t decreasing meniscus velocity, $\sim 0.21/\sqrt{t \, \mathrm{[ns]}}$ per nanosecond expressed here in terms of the tetracosane all-trans length of 3 nm. This rate falls below typical crystallization speeds of n-alkanes of 0.1 tetracosane length per ns [116] already after a few nanoseconds of elapsed imbibition time. Therefore, it is plausibel that the fast crystallization kinetics can heal out any occurring defects on t-scales much smaller than detectable by spontaneous imbibition dynamics.

Surface freezing at bulk n-alkane surfaces occurs over a small T-range ΔT of a few degrees centigrade (see Fig. 8.2). But the freezing temperature of the mesopore-confined liquid T_m is expected to be significantly shifted downwards [117]: $T_m < T_f$. X-ray diffraction experiments on n-tetracosane in porous Vycor® [118, 119] indicate freezing of the mesopore-confined liquid at $T_m \sim 40 \, °\mathrm{C}$, only. For that reason the evidence of an unchanged T_s despite the sizeable downward-shift of freezing temperature suggests that mesopore-confinement enables one to establish surface freezing over a much larger T-range than known of any other free surface of bulk n-alkane melts, i.e. $\Delta T \sim 16 \, °\mathrm{C}$.

I did not find unambigious hints of the Bragg peak pattern typical of the surface frozen state in supplementary conducted diffraction experiments. The dominating 70 % volume fraction of silica along with the coincidence of the first maximum of the structure factor of both the silica and the mesopore-confined liquid with the position of the dominating surface freezing Bragg peak render its detection in Vycor® particularly challenging. Fortunately, an alternative experiment delivered the first convincing and encouraging results. X-ray scattering experiments were performed on n-heptadecane (n-$C_{17}H_{36}$, another n-alkane that exhibits surface freezing in the bulk state) in an array of parallel-aligned tubular mesopores in a monolithic mesoporous silicon membrane that is porous silicon (see section 3.2 for details). The channels have a corona of ~ 1 nm thickness of native oxide. Because of its mainly single-crystalline silicon nature (see Fig. B.4), this matrix exhibits a low X-ray scattering background in the q-region of interest. Moreover, it allows one to probe different wave vector orientations with regard to the orientation of the ensemble of aligned nanochannels.

Plotted in Fig. 8.6 is the X-ray intensity as a function of the wave vector transfer perpendicular to the long channel axis ($q_\perp$ scans). There is a weak, but clearly visible Bragg peak at $q_{(10)}$, typical of the hexagonal surface backbone alignment

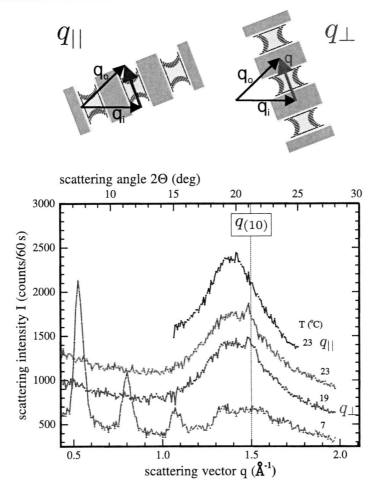

Figure 8.6.: X-ray diffraction patterns of a tubular silica nanochannel array in a mesoporous silicon membrane ($\sim 10\,\text{nm}$ mean channel diameter, $\sim 50\,\mu\text{m}$ length) 50 % filled with n-heptadecane with wave vector transfer parallel ($q_{\parallel}$) and perpendicular ($q_{\perp}$) to the long channel axis (see illustration of the scattering kinematics in the upper panel) for selected temperatures. The dotted line indicates the position of the (10) Bragg peak characteristic of the 2D-hexagonal backbone alignment of the surface frozen state in n-alkanes. It is $T_{\text{m}} = 13\,^{\circ}\text{C}$, $T_{\text{f}} = 21\,^{\circ}\text{C}$ and $T_{\text{s}} = 23\,^{\circ}\text{C}$. The pattern at $T = 7\,^{\circ}\text{C}$ is representative of the pore-solidified n-heptadecane. Courtesy of Anke Henschel, Saarland University, Saarbrücken, Germany.

of the surface frozen state, sitting on top of the broad maximum of the structure factor of the mesopore-confined liquid. It can be observed in the entire T-range from a few degree above bulk freezing T_{f} all the way down to freezing in the pores T_{m} and it vanishes along with the signature of the confined liquid upon pore crystallization (see diffraction pattern at $7\,^{\circ}\text{C}$). For this matrix/n-alkane combination the pore freezing point is $8\,^{\circ}\text{C}$ below bulk freezing, which means an existence range of the surface frozen layer of $10\,°C$ in comparison to just $2\,°C$ for the surface frozen state at the bulk melt surface.

Interestingly, and consistent with the picture of the surface frozen menisci, this peak is absent in all scans for which q is oriented parallel to the long axis of the channels ($q_{||}$, see Fig. 8.6 for a selected $q_{||}$-scan). This indicates that the molecules in the surface frozen layer are indeed arranged with their long, extended axes parallel to the long axes of the channels, as envisioned in the insets of Fig. 8.5. As a result, these cross-checking experiments on stable mesoscopic menisci in mesoporous silicon further corroborate the meniscus-frozen state in the restricted geometries of silica mesopores.

8.1.4. Conclusion

This study reveals the first evidence of surface freezing occurring at advancing menisci in silica mesopores. This observation testifies to a remarkable robustness of this archetypical alignment transition of the surface molecules both upon the mesoscale spatial confinement and upon the self-propelled movement of the interface in a rather complex, tortuous pore network. The influence of surface freezing on the imbibition flow dynamics can entirely be accounted for by the change in surface tension typical of surface freezing and does not additionally slow-down the meniscus movement. This discovery supports an impressive flexibility of the surface frozen layer similar to what was inferred from capillary wave spectroscopy [120, 121].

The surface frozen layer can be considered as a peculiar case of a surfactant. As a consequence, this study exemplifies how the formation of a surfactant layer at mesoscopic menisci can markedly affect the imbibition dynamics in nanocapillaries, similar to what has been ascertained for macropores [122, 123]. My study also suggests that mesopore confinement permits the establishment of the surface frozen state over much larger T-ranges than feasible at bulk surfaces.

8.2. Mesophases in 8OCB

Finally, the phase transition behavior of a thermotropic liquid crystal in mesopore confinement was examined. For this purpose I applied one of the best scientifically studied liquid crystals, the rod-like octyloxycyanobiphenyl (8OCB, see Fig. 8.7 for an illustration of its molecular structure). At $\sim 80\,°C$ it undergoes a transition from the isotropic to a nematic phase, which is accompanied by the occurrence of a characteristic shear viscosity minimum. Again, this distinctive behavior must affect the rise dynamics which enables one to examine the phase transition in mesopore confinement. Though, beforehand I will give a brief introduction of the physics of liquid crystals based on a tutorial of James A. Rego, Cal Poly Pomona, CA, USA.

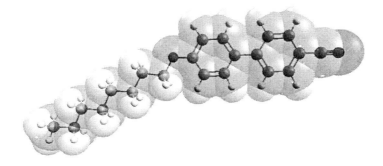

Figure 8.7.: Illustration of the rod-like molecular structure of the liquid crystal octyloxycyanobiphenyl (8OCB) with the phase transition behavior: crystalline $\xrightarrow{54\,^\circ\text{C}}$ A $\xrightarrow{67\,^\circ\text{C}}$ N $\xrightarrow{80\,^\circ\text{C}}$ I. The molecule's head (comprising the cyano and the biphenyl group) and its tail (the octyl chain) each have a length of slightly more than 1 nm. Its overall length is ~ 2 nm. As opposed to the hydrocarbons the 8OCB molecule is rather rigid.

8.2.1. Properties of Liquid Crystals

Basically, liquid crystals (LCs) are anisotropic liquids. They possess the fluidity of a true liquid, as well as varying degrees of long range order normally associated with crystalline solids. The anisotropy of the bulk phase has its origins in the anisotropy of the molecules comprising the fluid itself. This general anisotropy of shape can have various manifestations, e.g., rods, disks, bananas, and so forth. Anyhow, this intrinsic constitution of the liquid gives rise to the alignment of the molecules and, therefore, to the anisotropic properties of the liquid crystal.

Liquid crystal phases are called mesophases. Thermotropic liquid crystals exhibit one or more anisotropic liquid phases between the melting point T_f and the temperature at that they become isotropic T_c; this temperature is called the clearing point due to the milky, translucent appearance of liquid crystal phases. These phases generally occur reproducible with heating or cooling.

The most common phases are illustrated in Fig. 8.8. In the isotropic (I) state the rod-shaped molecules are randomly oriented and the molecular centers of mass move as in any liquid. Upon cooling the molecules, at a certain well defined temperature, often self-reorient with their long axes parallel to each other whereas the centers of mass are still isotropically distributed. This phase is called the nematic phase (N). For an isotropic liquid, averaging molecular orientations gives no result, since there are as many molecules lying along one axis as along another. In the nematic phase, averaging molecular orientations gives a definite preferred direction, which is referred to as the director $\vec{n}$. However, it is important to remember that liquid crystals are liquids, meaning that, although there is an average order, molecules are constantly flowing and moving, changing position and orientation.

Upon further cooling from the nematic phase, the molecules will often self-

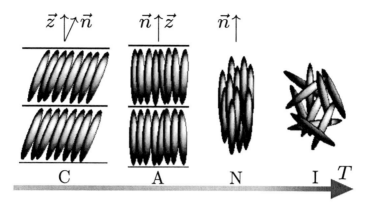

Figure 8.8.: Illustration of common liquid crystal phases. Upon heating a frozen liquid crystal one might pass the following mesophases: smectic C (C), smectic A (A), and nematic (N). Beyond the clearing point T_c the liquid crystal is in the isotropic state (I).

assemble into layers. These layered phases are called smectic liquid crystals. When the director is parallel to the layer normal $\vec{z}$, the phase is referred to as a smectic A (A). When cooling from the A phase, a more ordered smectic phase is often seen, in which the molecules are tilted with respect to the layers. This tilted, layered phase is called the smectic C (C).

8.2.2. Shear Viscosity Minimum and Presmectic Divergence of Flowing Nematic Liquid Crystals

The momentum transport in nematic liquid crystals shows an anisotropy since it depends on the mutual orientations of the macroscopic molecular alignment (the director $\vec{n}$), the flow velocity ($\vec{v}$) and the velocity gradient (∇v). In 1946 Miesowicz defined three principal shear viscosity coefficients of nematics [124], which can be measured in three different Couette flow experiments sketched in Fig. 8.9. Typically magnetic fields are applied in order to align the molecules in the nematic sample. Intuition suggests that the lowest resistance to the nematic flow, i.e. the lowest viscosity value, should be η_2. Among the two remaining viscosities, η_1 should have the highest value.

Indeed, when the orienting magnetic field, i.e. the director $\vec{n}$, is parallel to the velocity $\vec{v}$ of the nematic flow, the lowest viscosity is recorded (see open symbols in Fig. 8.10). Nevertheless, this relatively simple picture of the viscosity of nematic liquid crystals is disturbed for the compounds exhibiting the transition to the smectic A phase. Then, with decreasing temperature, the viscosity η_2 shows a strong increase and goes to infinity at the N-A phase transition. The η_1 and η_3 viscosities are almost unaffected. At a temperature that is a few degrees below the temperature at that the transition to the smectic A phase takes place, the viscosities η_2 and η_3 interchange their roles and then the lowest nematic viscosity corresponds to the flow in the η_3 configuration. The presmectic behaviour of the

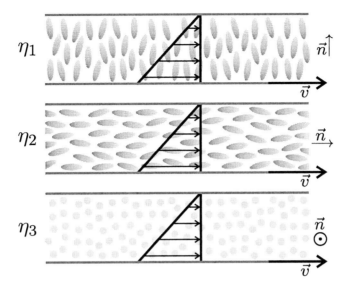

Figure 8.9.: The experimental Couette flow conditions for measurements of the three Miesowicz shear viscosity coefficients of nematic liquid crystals: η_1 for $\vec{n} \perp \vec{v}$ and $\vec{n} \| \nabla v$, η_2 for $\vec{n} \| \vec{v}$ and $\vec{n} \perp \nabla v$, η_3 for $\vec{n} \perp \vec{v}$ and $\vec{n} \perp \nabla v$.

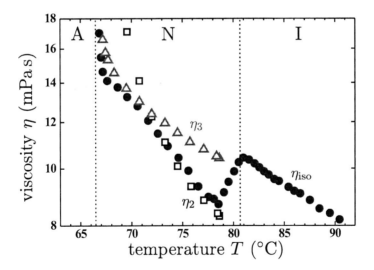

Figure 8.10.: Miesowicz shear viscosities η_2 ($\square$) and η_3 ($\triangle$) of the liquid crystal 8OCB compared to its free flow viscosity η_{iso} ($\bullet$) according to [125].

η_2 viscosity is due to the formation of precursors of smectic planes with $\vec{z} \| \vec{v}$ that would be immediately destroyed by the velocity gradient, thus this configuration is rendered unfavorable.

The behavior of the freely flowing compound obeys a general principle that can be formulated as follows: a free fluid adopts such a manner of flow, as corresponds to the minimum of its viscosity at given conditions [126]. Accordingly, the transition from the isotropic to the nematic phase manifests itself in a strong decrease of the shear viscosity η_{iso} that is very close to η_2. Consistently, beyond the presmectic cross-over of η_2 and η_3 the viscosity of the freely flowing liquid crystal η_{iso} follows η_3 (see filled symbols in Fig. 8.10). This result is interpreted in terms of rearrangements of the molecular alignment $\vec{n}$ with respect to the velocity field $\vec{v}$, which can easily be assessed by means of examinations of the compound's viscosity.

8.2.3. Results

This distinctive behavior of the viscosity must be mapped in the T-dependence of the measured mass uptake. In Fig. 8.11 some representative measurements of 8OCB invading V5 are shown along with their corresponding $\sqrt{t}$-fits. In analogy to the proceeding at the evaluation of the surface freezing transition I conducted a quantitative analysis of the phase transition behavior of 8OCB in mesopore confinement referring to the prefactors obtained from the fits.

Again, the prefactors are normalized by the prefactor value at an arbitrarily chosen temperature T^{n}. The thereby obtained normalized imbibition speeds $v^{\text{n}}(T)$ calculated from the fitting parameters are indicated by the single points in the upper panel of Fig. 8.12. Assuming the $\eta_{\text{iso}}(T)$ behavior according to Fig. 8.10 along with the T dependency of σ and ρ (corresponding to Tab. A.2) one can calculate the theoretical values of $v^{\text{n}}(T)$ based on Eq. (8.3). They are plotted as a solid line in the upper panel of Fig. 8.5.

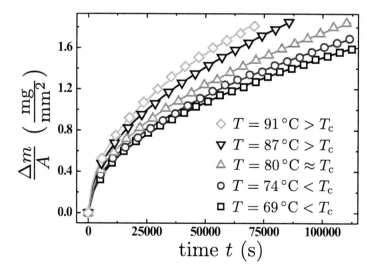

Figure 8.11.: Specific mass uptake of V5 due to the imbibition of the liquid crystal 8OCB as a function of the time for selected temperatures below and above the clearing point $T_{\text{c}} \approx 80\,^{\circ}\text{C}$. Solid lines correspond to $\sqrt{t}$-fits. The data density is reduced by a factor of 2500.

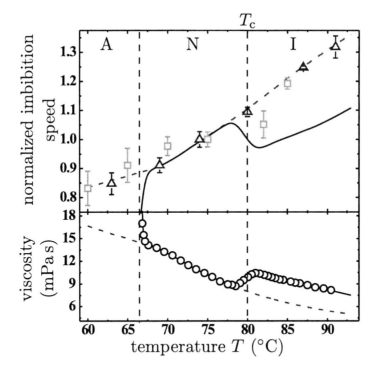

Figure 8.12.: (upper panel): Measured normalized imbibition speeds v^n (for $T^n = 75\,°C$) of 8OCB in V5 ($\triangle$) and V10 ($\square$), respectively, in comparison with values calculated on the basis of the viscosity values in the lower panel (—). (lower panel): T-dependent viscosity of 8OCB by way of comparison (see Fig. 8.10 for a detailed view). The dashed lines correspond to extrapolations of the calculated imbibition speeds and of the viscosity in the absence of the shear viscosity minimum and the presmectic divergence, respectively.

8.2.4. Discussion

The results in Fig. 8.12 show a variety of astonishing features. First of all, both V5 and V10 reveal comparable characteristics of the invasion dynamics of the liquid crystal 8OCB. Anyhow, only in the nematic phase the T-dependent behavior of the measured imbibition speeds coincides with the prediction based on the η_{iso} values presented in Fig. 8.10. In particular, the distinctive bump in the proximity of the clearing point T_c as the direct manifestation of the shear viscosity minimum in the theoretical behavior of v^n is unambiguously absent. The imbibition speed rather increases monotonously with increasing temperature.

As mentioned before this distinctive feature of the viscosity at the clearing point is caused by the inset of an alignment of the molecules (in the nematic phase) with respect to the flow direction, that is $\vec{n} \,||\, \vec{v}$. Therefore its absence can intuitively be interpreted in terms of an already existent alignment of the molecules beyond T_c. It is obvious to conclude that such an alignment is easily induced by the extreme spatial confinement to cylindrical pores with diameters that are not more than five times the length of the molecule itself. In addition, considering their rigidity and, in particular, the parabolic flow profile established in the pore, it is hard to think of any alternative to the tendency of molecular alignment parallel to the pore axis and, consequently, to the flow velocity $\vec{v}$. From this point of view the absence of the shear viscosity minimum and, consistently, of the nematic to isotropic phase transition is not surprising at all but rather consequent. In this context it is more suitable to label the phase beyond the clearing point not isotropic but *paranematic* (P). This term preeminently discloses the solely confinement-induced alignment of the liquid crystal.

It was demonstrated experimentally [127–131], in agreement with expectations from theory [132–134], that there is no 'true' I-N transition for liquid crystals confined in geometries spatially restricted in at least one direction to a few nanometers. The anchoring at the confining walls, quantified by a surface field, imposes a partially orientational, that is, a partially nematic alignment of the confined liquid crystal, even at temperatures T far above the clearing point T_c. The symmetry breaking does not occur spontaneously, as characteristic of a genuine phase transition, but is enforced over relevant distances by the interaction with the walls.

The theory of such a paranemtic phase is enhanced by recent birefringence measurements of liquid crystals confined to an array of parallel, nontortuous channels of 10 nm mean diameter and 300 µm length in a monolithic silica membrane (see section 3.2 and Fig. 3.5 for details) [41]. These measurements elucidate that the surface anchoring fields render the bulk discontinuous I-N transition to a continuous P-N transition (see Fig. 8.13). The transition temperature T_c is found to be changed only marginally, due to a balance of its molecular alignment induced upward and its quenched disorder (attributable to wall irregularities) induced downward shift. This agrees with the observations of liquid crystals imbibed in tortuous pore networks [127, 128].

Interestingly, due to the complete absence of the shear viscosity minimum in the results of the T-dependent series of imbibition measurements shown in Fig. 8.12 a definition of a P-N transition temperature in confinement is not possible at all. This is confirmed by a simple extrapolation of the η_2 viscosity to higher T, which saliently reproduces the measured imbibition speeds beyond the bulk clearing point. This can only be interpreted in terms of an extremely high degree of orientation already existent in the paranematic phase; at least higher than suggested by the birefringence measurements in Fig. 8.13. This difference is most probably due to the basically differing detection method of the P-N transition: normally birefringence or calorimetry (DSC) measurements performed with the confined *static* liquid are applied for this purpose. But, the viscosity measurements presented here refer to the liquid's dynamics in mesopore confinement. As already mentioned before, the additional emerging flow velocity and in particular the velocity gradient seemingly enhance the paranematic orientational alignment in the η_2-configuration significantly.

The second remarkable feature of the results presented in Fig. 8.12 is the absence

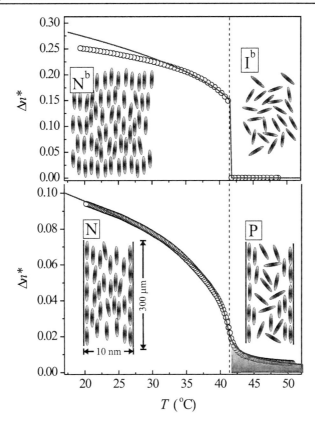

Figure 8.13.: Optical birefringence of the liquid crystal 7CB measured in the bulk state (upper panel) and in the silica nanochannels (lower panel) as a function of the temperature T in comparison to fits (solid lines) based on a model discussed in Ref. [41]. The finite birefringence characteristic of the paranematic phase is shaded down to the P-N transition temperature. The dashed line marks the bulk clearing point. As insets in the upper and lower panel, the bulk isotropic (I^b) as well as the bulk nematic (N^b) phases upon homeotropic alignment, and the confined paranematic (P) and nematic (N) phases are illustrated, respectively. Courtesy of Patrick Huber, Saarland University, Saarbrücken, Germany.

of the presmectic divergence of the viscosity of the freely flowing liquid crystal. This would result in a dramatic drop of the imbibition speed due to the inset of smectic layering. However, the measured values do not indicate such an effect. Its absence rather suggests a suppression of the A phase in favor of the N phase. This is elucidated by a simple extrapolation of the viscosity to lower T as indicated in Fig. 8.12, which preeminently reproduces the measured imbibition speeds.

What are the reasons for this discovery? First of all, the confinement to a cylindrical pore (rather than to a film geometry like in the Couette flow depicted in Fig. 8.9) renders the η_3 viscosity as unfavorable as the η_1 viscosity. This again clarifies the high stability of the nematic η_2-configuration in confinement. A cross-over behavior as ascertained for the freely flowing bulk liquid can hence be excluded. Yet, the presmectic divergence of η_2 is caused by the destruction of precursors of smectic planes for $\vec{n} \parallel \vec{v}$. From this point of view the suppression of the A phase is not surprising at all but a mere consequence of the overall stabilization of the η_2-configuration in the mesopore confinement.

Even the static mesopore-confined liquid crystal shows such modified mesophase behavior [129]. For example the heat capacity anomaly typical of the second-order N-A transition in rod-like liquid crystals immersed in aerogels is absent or greatly broadened [135, 136]. Nuclear magnetic resonance (NMR) measurements revealed the lack of pretransitional smectic layering due to the rough surface of the confining walls [137]. Furthermore, a systematic study of the influence of the degree of confinement indicates that the N-A transition becomes progressively suppressed with decreasing pore radius whereas the stability range of the nematic phase is increased [138].

Finally, I will present some results from a more quantitative analysis of the measurements. According to the evaluation method used in chapter 6 one arrives at slip lengths of $b = (-1.11 \pm 0.23)$ nm for V5 and $b = (-1.54 \pm 0.31)$ nm for V10. Again, a sticking layer boundary condition has to be applied. Apparent velocity slippage at the walls, as was expected to be associated with the molecular alignments in the channels [139], could not be detected.

8.2.5. Conclusion

In conclusion the T-dependent imbibition measurements of the liquid crystal 8OCB revealed that confinement plays a similar role as an external magnetic field for a spin system: the strong first-order I-N transition is replaced by a weak first-order or continuous paranematic to nematic transition, depending on the strength of the surface orientational field [140]. Based on detailed knowledge of the static (equilibrated) liquid's behavior in the mesopores as deduced from previously conducted birefringence experiments, I was able to procure complementary results with respect to its dynamic (non-equilibrium) behavior. The additional emerging flow velocity and in particular the velocity gradient enhance the paranematic orientational alignment significantly rendering the P-N transition even broader than known from the equilibrium state. Due to the high stabilization of the η_2-configuration in the N phase the A phase is suppressed and the stability range of the nematic phase is increased. Nevertheless, despite the molecular alignment no indications of velocity slip were found.

Part III.

Forced Throughput Study

The measurements that will be presented in Part III of this thesis aim at the flow behavior of liquids in networks of mesopores. From this point of view they are complementary to the study of the rise dynamics already presented in chapter 6. However, in the following investigation the liquid flow is driven by an external force contrary to spontaneous (self-propelled) imbibition. This additionally renders possible the examination of non-wetting fluids.

9. Fundamentals & Experimental Setup

The capillary rise dynamics discussed in Part II sensitively depend on the liquids' surface tension σ and the dynamic contact angle θ_D. This provides the opportunity for an examination of these properties in confinement, e.g., the observation of the surface freezing transition presented in section 8.1. But, because of this fact the spontaneous imbibition measurements are subject to a general limitation: only liquids that wet the porous substrate can be examined. In order to get rid of this constraint a forced throughput method, sometimes referred to as forced imbibition, was applied. Here, the liquid does not self-propelled (or spontaneously) invade the porous host; the flow is rather driven by an external force.

9.1. Basics

The dynamics of the flow of a liquid through a porous host of thickness d and cross-sectional area A (that is already completely filled with the liquid) can directly be related to Darcy's law Eq. (2.4) in conjunction with the permeability K according to Eq. (2.7). For a pore network with mean pore radius r_0, porosity ϕ_0 and tortuosity τ and with the liquid's viscosity η, in terms of the volume flow rate this finally reads

$$\dot{V} = \underbrace{\frac{A\phi_0}{8\,d\eta\tau}\frac{r_h^4}{r_0^2}}_{C_V}\,\Delta p \qquad (9.1)$$

with Δp denoting the (externally generated) pressure drop that is applied along d. By determining the prefactor C_V through a measurement of $\dot{V}(\Delta p)$ the hydrodynamic pore radius r_h is easily accessible. This principle has already been used by Peter Debye and Robert Cleland in their seminal work on the flow of hydrocarbons through porous Vycor® glass [4].

The measuring principle just stated is rather simple and not subject to as many hypotheses as there are for the spontaneous imbibition measurements. Anyhow, the implementation of the measuring method entails certain demands, which prevent a general applicability. For one thing the total volume flow rates through the Vycor® samples are on the order of a few nanoliters per second only. They can be measured as small changes in the capacitance of a calibrated cylindrical capacitor. Consequently, the resolution is directly proportional to the liquid's dielectric constant ε. According to this precondition, hydrocarbons with their relatively low $\varepsilon \approx 2$ are only moderately applicable whereas water is particularly suitable because of its permittivity being approximately 50 times larger (see Fig. 9.3).

For another thing, the experimental setup provides the opportunity to conduct experiments up to 80 °C. Yet, the assembly of the setup can only be accomplished with a compound in the liquid state. Hence, no substance with a melting point beyond room temperature can be applied. Moreover, for the complete filling of the reservoir and the supply channels one requires a considerable amount of the liquid. This further restricts the group of applicable liquids. Eventually such forced throughput measurements were carried out with water and the linear hydrocarbon n-hexane ($n\text{-}C_6H_{14}$).

9.2. Experimental Setup & Measuring Principle

The experimental setup for the forced throughput measurements, the membrane flow apparatus (MFA), is illustrated in Fig. 9.1. All parts liquid containing (see Fig. 9.2 for a detailed view) are immersed in a water bath, which can be heated up to 80 °C. The external force is provided by a highly pressurized gas. For this purpose the flow system is connected to a gas handling, which supplies the gas

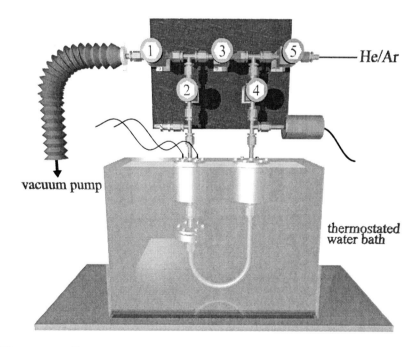

Figure 9.1.: Raytracing illustration of the membrane flow apparatus (MFA) consisting of a gas handling and the actual flow system. The latter is temperature-controlled in a water bath (for a more detailed view see Fig. 9.2). The gas handling can be evacuated by a vacuum pump and can be filled with helium or argon gas. The setup was partly constructed in the course of the diploma thesis of Stefanie Greulich [141].

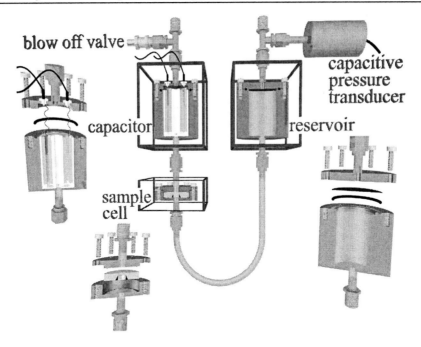

blow off valve

capacitive pressure transducer

capacitor

reservoir

sample cell

Figure 9.2.: Exploded view of the membrane flow apparatus (raytracing illustration). The liquid is pressed from the reservoir on the high pressure side through the membrane in the sample cell into a cylinder capacitor on the low pressure side. Flow rates were measured via capacitance changes. The pressure applied by either the helium or the argon gas can be measured employing a capacitive pressure transducer ($p_{\mathrm{max}} = 200\,\mathrm{bar}$). Beyond the capacitor the pressure's upper limit is fixed to 1 bar by a blow off valve.

via valve 5 and 4. The valves 1, 2, and 3 permit an initial evacuation of the handling; during the measurements they are normally closed thereby separating the (right) high pressure from the (left) low pressure side. The complete setup is manufactured inhouse and made of stainless steel. This allows for maximum pressures of up to 70 bar, which can be measured with a capacitive pressure transducer. The pressure beyond the capacitor is fixed to the upper limit of 1 bar by means of a blow off valve.

For most measurements the liquid was pressurized with high purity helium gas (6.0). This choice was made in order to lessen the impact of a major flaw in the measuring method: the liquid stands in direct contact with the highly pressurized gas. As a consequence, it cannot be avoided that a significant amount of gas is dissolved in the liquid and thereby considerably influences the measuring conditions. Some imaginable consequences will be discussed in the next chapter. However, with the usage of an inert gas at least chemical reactions can be prevented. What is more, helium is the gas that is, at room temperature, the least soluble in water [142]. In order to study a possible influence of the solubility of the gas on the dynamics argon (purity 5.7) can be used as well. Furthermore, the liquid in the reservoir can be separated from the gas using a flexible membrane. Yet, finding a membrane with an appropriate chemical constitution that meets all required conditions is rather challenging. This will be discussed in the next chapter.

Via the supply channel the pressurized liquid in the reservoir reaches the cell with the cylindrical sample of typically $d = 4\,\mathrm{mm}$ thickness and a diameter of $6\,\mathrm{mm}$. The latter is thoroughly glued into a copper sample holder using the two component, thermally conductive epoxy encapsulant Stycast with the catalyst 24LV from Emerson & Cuming. With this procedure one must not only accomplish the task of fixing the sample, but also that of sealing the sample's side facets in order to guarantee the flow through the top and bottom facets only. Or equivalently: the procedure should ensure that the pressure drop is applied along the complete sample thickness d.

Beyond the sample cell the cylindrical capacitor is attached. Due to the liquid flow through the sample the liquid level in the capacitor rises thereby changing the capacitance. The latter can accurately be ascertained employing a multi frequency LCR meter (HP 4275A) at the frequency $f = 500\,\mathrm{kHz}$. This value was chosen with regard to water's high dielectric loss within the microwave range (roughly between $1\,\mathrm{GhZ}$ and $1\,\mathrm{THz}$, e.g., microwave oven $f = 2.45\,\mathrm{GHz}$), which would entail additional inaccuracies due to the strong f dependency of the permittivity. For $f = 500\,\mathrm{kHz}$ the dielectric constant only shows the persistent dependence on the temperature T.

9.2.1. Calibration

For a direct relation between the shift in the capacitance C and the related change in the liquid volume V in the capacitor the latter was calibrated. For this purpose its capacitance was measured while it was stepwise filled with specific amounts of the respective liquid. Since the permittivity is a function of the temperature this procedure was performed for all relevant T. In general, each calibration was repeated at least five times. Some of the resultant $C(V)$ curves are exemplarily shown in Fig. 9.3.

The plots confirm the before-mentioned good applicability of water because of its high dielectric constant as compared with n-hexane. Additionally, the influence

Table 9.1.: Calibration factors C_{cal} (in $\frac{\mathrm{pF}}{\mathrm{ml}}$) of water and n-hexane at relevant temperatures T (in °C).

T	water	n-hexane
25	96.9 ± 0.2	1.198 ± 0.004
40	91.2 ± 0.1	1.181 ± 0.008
50		1.197 ± 0.011
60	84.9 ± 0.2	

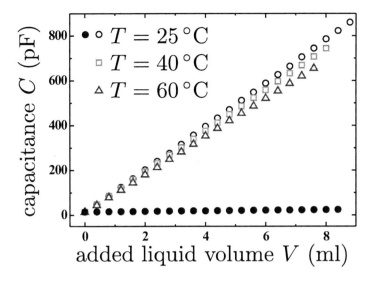

Figure 9.3.: Calibration measurements of the cylindrical capacitor for water (open symbols) and n-hexane (filled symbols) at selected temperatures. The capacitance C was measured as a function of the liquid amount V filled into the capacitor. For the empty capacitor it is $C \approx 15\,\mathrm{pF}$.

of the temperature is clearly recognizable: with increasing T the polarizability decreases due to the enhanced microscopic mobility of the molecules. Macroscopically this behavior is expressed in terms of a decreasing permittivity of the liquid.

One is now able to connect a certain change in C with an equivalent change in V via a calibration factor C_{cal} that is the slope of the shown calibration curves: $\frac{\mathrm{d}C}{\mathrm{d}V} \equiv C_{\mathrm{cal}}$. For the liquids applied and the relevant temperatures the respective values are listed in Tab. 9.1. The flow of n-hexane was measured at $50\,^{\circ}\mathrm{C}$ instead of $60\,^{\circ}\mathrm{C}$ (as for water) because of the increasing noise in the proximity of its boiling point at $69\,^{\circ}\mathrm{C}$.

Using Eq. (9.1) this finally results in a relationship between the measured variation of the capacitance C as a function of the time t (at a given applied pressure gradient Δp) and the flow dynamics in confinement

$$\dot{C} = C_{\mathrm{cal}}\dot{V} = C_{\mathrm{cal}}\,C_{\mathrm{V}}\,\Delta p \tag{9.2}$$

expressed in terms of the prefactor C_{V} (see Eq. (9.1)). The most accurate way to deduce C_{V} is extracting the slope of a $\dot{V}(\Delta p) = \frac{\dot{C}(\Delta p)}{C_{\mathrm{cal}}}$ plot.

10. Flow Dynamics in Porous Vycor®

This chapter centers on a forced throughput study of water and the linear hydrocarbon n-hexane in porous Vycor® glass. For one thing such measurements aim at the flow behavior of liquids in networks of mesopores and therefore are complementary to the study of the rise dynamics already presented in chapter 6. Nevertheless, the measuring method enables one to examine non-wetting fluids as well. For this purpose the silanization procedure presented in section 3.1 was applied.

10.1. Dynamics in Untreated Porous Vycor®

In this first part I will present results obtained from forced throughput measurements on untreated Vycor®. The raw data signal of the capacitance change C as a function of the time t is exemplarily shown in Fig. 10.1 for the flow of water in V10 at $T = 25\,°C$ and for selected applied pressures generated with helium

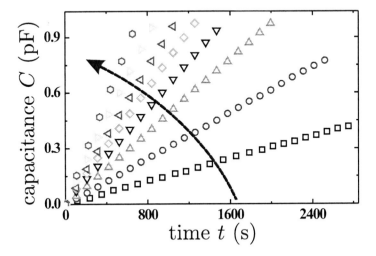

Figure 10.1.: Time dependent variation in the capacitance C of the cylinder capacitor due to the flow of water through V10 at 25 °C for a series of applied external pressures. The arrow indicates the direction of increasing Δp. The shown measurements correspond to: 8 bar, 16 bar, 24 bar, 31 bar, 37 bar, 45 bar, 54 bar, and 70 bar. The data density is reduced by a factor of 20.

gas. It is evident that with increasing Δp the variation in C with t, that is the slope $\dot{C}$, increases gradually. This result can directly be interpreted in terms of an increasing volume flow rate $\dot{V} = \frac{\dot{C}}{C_{\text{cal}}}$ with increasing pressure.

10.1.1. Results

In Fig. 10.2 some of the resultant volume flow rates $\dot{V}$ of water in both V5 and V10 at three different temperatures are plotted as a function of the applied external pressure Δp. The same was done for the flow of n-hexane. Some of the corresponding results are shown in Fig. 10.3. However, due to the rather low calibration factor of n-hexane as compared with that of water, the measuring time had to be increased in order to gain a proper signal with sufficient resolution. For that reason the overall data density is markedly reduced for n-hexane.

In principle all data sets show a linear relation compliant with Eq. (9.2). The comparison between different temperatures implies – at least for water – a distinct T dependence of the proportionality constant C_V: the latter increases with increasing temperature. According to Eq. (9.1) this behavior is solely determined by the temperature dependence of the liquid's viscosity (see Tab. A.1 and Tab. A.4). Qualitatively this is true: at higher temperatures the lower viscosities cause an increase in C_V. But, in the temperature region of interest the T dependency of

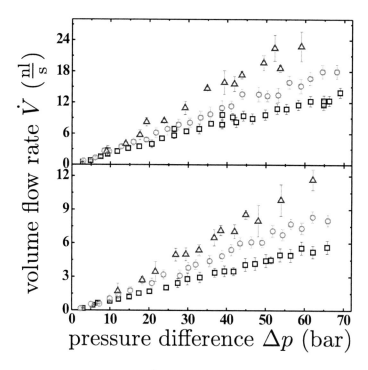

Figure 10.2.: Volume flow rates $\dot{V}$ of water in V10 (upper panel) and V5 (lower panel) as a function of the applied external pressure difference Δp at three different temperatures: 25 °C ($\square$), 40 °C ($\circ$), and 60 °C ($\triangle$).

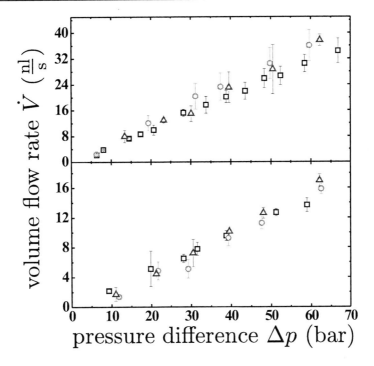

Figure 10.3.: Volume flow rates $\dot{V}$ of n-hexane in V10 (upper panel) and V5 (lower panel) as a function of the applied external pressure difference Δp at three different temperatures: 25 °C ($\square$), 40 °C ($\bigcirc$), and 50 °C ($\triangle$).

the viscosity of water is more distinctive than that of n-hexane. This behavior renders the effect more distinct for water.

In a subsequent step the values of the hydrodynamic pore radii r_h may be calculated from the extracted slopes C_V. Based on the matrix properties stated earlier (see Tab. 3.1) and on the known sample dimensions A and d one arrives at the slip lengths $b = r_h - r_0$ listed in Tab. 10.1. Since these values should principally be independent of the respective liquid and the measuring temperature they allow for a more quantitative analysis and comparability of the results.

10.1.2. Discussion

First of all, nearly all extracted slip lengths are again negative suggesting a sticking layer boundary condition in compliance with the results from the imbibition study. Anyhow, by means of capillary rise measurements the thickness of the sticky layer was found to be approximately 5 Å for water as well as hydrocarbons in both V5 and V10. But the values stated in Tab. 10.1 all deviate slightly towards lower values and eventually b turns even positive for water in V10. For water there seems also to be a marginal increase in b with T whereas there is no systematic dependency for n-hexane. Contrasting the results for water with the results for n-hexane it turns out that the slip lengths for water are always higher

Table 10.1.: Slip lengths b (in Å) of water and n-hexane flowing through V5 and V10, respectively, as extracted from forced throughput measurements at three different temperatures T (in °C). The liquids were pressurized with helium.

T	V5		V10	
	water	n-hexane	water	n-hexane
25	-2.6 ± 1.6	-3.8 ± 1.8	-0.1 ± 2.2	-2.8 ± 2.6
40	-1.8 ± 2.1	-4.5 ± 2.2	-0.3 ± 2.9	-1.9 ± 2.9
50		-3.5 ± 2.4		-3.2 ± 2.8
60	-1.3 ± 2.3		0.7 ± 3.6	

than those for the alkane. Furthermore, the values for V10 are systematically increased as compared with V5.

The bottom line of these results is that the forced imbibition dynamics are generally increased as compared to the spontaneous imbibition. Additionally, there are configurations regarding the flowing liquid and the substrate that seemingly facilitate higher slip lengths. This observation can be condensed as follows:

$$b(\text{hexane}) \ < \ b(\text{water})$$
$$b(\text{V5}) \ < \ b(\text{V10}) .$$

The increase of b with increasing temperature for water is only vague but should not remain unmentioned at this point.

To sum up, the forced throughput measurements display rather peculiar results, which do not show the consistency of the imbibition study presented earlier. Nonetheless, it is possible to consistently trace the overall increased dynamics and their curious depencencies obtained in this study back to a basic difference in the measuring conditions. Basically there are two such differences that might fundamentally cause variations in the measured dynamics.

First, in the present study the liquid flow is generated by a positive pressure Δp whereas in the imbibition configuration the driving force is generated by the negative Laplace pressure $-p_{\mathrm{L}}$. Yet, the changes in the viscosity and the density at the pressures probed in the experiments entail deviations on the order of maximum 3 % only [94, 95]. Moreover, one would expect a far more distinctive and to the marginal behavior contradictory T dependence of this effect as achieved in the experiments. Eventually this hypothesis must be ruled out.

Second, and this is the more promising issue, which has already been addressed: in the forced imbibition measurements the liquid stands in direct contact with the highly pressurized gas. Consequently it is unavoidable that a relevant amount of gas is dissolved in the liquid and thereby influences the measuring conditions significantly. To date it has often been reported that dissolved gas modulates slip [27, 28]. For Newtonian fluids enhanced dynamics were found to be consistent with a two-layer-fluid model, in which a layer < 1 nm thick, but with viscosity

10 - 20 times less than the bulk fluid, adjoins each solid surface [20]. A potential mechanism to explain the genesis of this layer was discussed by Olga Vinogradova [91] and formalized by Pierre-Gilles de Gennes [143], who hypothesized that shear may induce nucleation of vapor bubbles; once the nucleation barrier is exceeded the bubbles grow to cover the surface, and the liquid flow takes place over this thin gas film rather than the solid surface itself. Hence, the segregation of gas at the near-surface region seems to facilitate some kind of low-density surface excitations, but the nature of these is not understood well at this time.

SFA measurements on tetradecane performed by Granick *et al.* impressively elucidate this theory [27]. The experiments showed that whereas no-slip behavior was obeyed when the tetradecane had been saturated with carbon dioxide gas, massive deviations from this prediction were found when the tetradecane was saturated with argon. This makes it seem likely that if argon is segregated at the solid walls it creates a low-viscosity boundary layer, thus greasing the flow of fluid past that surface. Steve Granick made the conjecture that the amount of segregation is a materials property of the fluid, the chemical makeup of the surface, and the chemical identity of the dissolved gas. Argon possesses only low solubility in tetradecane what may have made it more prone to segregate at the surfaces. Interestingly, for water the saturation with *both* argon *and* carbon dioxide resulted in slip.

According to these results and considerations the shear rate and the solubility of the gas (hereinafter denoted as S) determine the possible influence of such segregation at a near-surface region. In the following I will assess whether a process like this can be responsible for the observed peculiarities.

First of all, an impact of the shear rate can indeed be noticed. Since for a given applied pressure difference the maximum shear rate in a channel increases with the fifth power of the channel radius [148], one may conclude that segregation, and therefore enhanced flow dynamics are more likely in V10 than in V5. This behavior is expressed by the higher slip lengths in V10 as compared to V5.

A potential effect caused by the gas' solubility in the respective liquid can be assessed considering the solubilities listed in Tab. 10.2. It is obvious that for a given temperature the solubility of helium is higher in n-hexane than in wa-

Table 10.2.: Solubilities S (in $\frac{mg}{\ell}$) of helium and argon in water [142, 144] and n-hexane [145–147] at 1 bar for selected temperatures (in °C).

T	helium		argon	
	water	n-hexane	water	n-hexane
20	1.45		62.0	
25		7.9	55.9	780
40	1.26	9.6	45.7	730
60	1.06		37.7	

Table 10.3.: Slip lengths b (in Å) of water and n-hexane flowing through V10 pressurized by two different gases, namely helium (He) and argon (Ar).

liquid	temperature	He	Ar
water	25 °C	-0.1 ± 2.2	-1.2 ± 3.3
n-hexane	25 °C	-2.8 ± 2.6	-2.3 ± 3.1
	40 °C	-1.9 ± 2.9	-1.6 ± 3.5

ter. According to the before-mentioned segregation of gas and the enhanced flow dynamics should be more likely for water than for n-hexane. This prediction coincides with the observed systematically higher slip lengths for water. Even the vague increase in b with the temperature T is consistent with the slight decrease in S with increasing temperature.

So far, Steve Granick's conjecture is in accord with the observed behavior. For an additional test some forced imbibition experiments in V10 were also carried out with argon instead of helium. According to Tab. 10.2 its solubility in water is about 40 times higher than that of helium; in n-hexane it is even up to 100 times higher. Accordingly, for both experiments one would expect smaller slip lengths as compared to the measurements with helium. However, the results of the investigation listed in Tab. 10.3 do not confirm this prediction unambiguously. For water there is indeed a slight decrease in b. Though, for n-hexane no distinct reduction can be observed.

At this point it is worthwhile considering a basic difference between the SFA measurements of Steve Granick and the present forced throughput study: the liquid in the reservoir is saturated with the respective gas at pressures up to 70 bar. Since the saturation concentration is nearly directly proportional to the applied pressure (see Tab. 10.4), the liquid at the high pressure side of the sample comprises up to 70 times more dissolved gas than could ever be stable on the low pressure side of the sample. Because of the linear pressure drop along the sample thickness d this necessarily leads to a gradual degassing of the liquid.

Eventually this phenomenon leads to a transport of gas from the high to the low pressure side. This effect is excellently illustrated in Fig. 10.4. Frequent drops in the measured capacitance C as a function of the time indicate the raise of bubbles in the capacitor and the consequent descent of the liquid level. According to the solubilities in Tab. 10.2 these drops are more frequent for measurements with the liquid hexane and the gas argon. What is more, the plot reveals a higher bubble

Table 10.4.: Solubilities S (in $\frac{mg}{\ell}$) of helium in water at 25 °C for selected pressures p (in bar) [144].

p	1	3	5	7.5	10	25	50	75
S	1.45	4.35	7.09	10.6	14.1	35.1	70.1	105

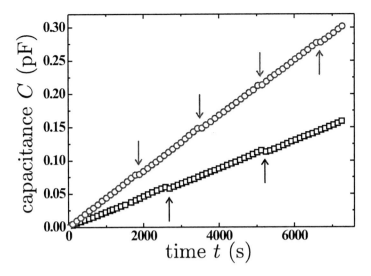

Figure 10.4.: Illustration of frequent drops (see arrows) in the raw data signal $C(t)$ of the forced imbibition measurements of n-hexane in V10 at 25 °C due to the raise of bubbles of translocating gas. The applied pressure differences are 35 bar ($\square$) and 68 bar ($\bigcirc$). The data density is reduced by a factor of 10.

rate at higher pressures. One might assume that the ever present degassing inevitably entails overall enhanced flow dynamics, which are slightly modulated by the before mentioned influences of the shear rate and the solubility of the gas.

10.1.3. Conclusion

As predicted from the spontaneous imbibition measurements here again a sticking layer boundary condition must be applied. One may conclude that up to a certain degree the peculiar underlying behavior of the slip length fits rather well into the model of slip modulated by dissolved gases. Nonetheless, there are still many open questions, which require additional investigations. In particular it will be necessary to conduct comparative experiments, for which the gas is separated from the liquid and for that reason an influence of dissolved gases can be excluded. Only by this means it will be feasible to unambiguously trace the enhanced dynamics back to this effect. The observations reported here can merely be seen as hints for such an influence.

First experiments with separating membranes constituted of polyisoprene failed due to its decomposition in contact with hexane. The second choice, polyurethane, is not affected by any of the liquids. Nevertheless, the diffusion rate of gases and in particular of helium through such a membrane is not negligible. Actually, no distinct influence of the used membrane could be found in the experiments. This point will be rather challenging in future experiments.

10.2. Dynamics in Silanized Porous Vycor®

The high significance of the liquid-substrate interaction in such extremely restricted geometries has been pointed out several times so far. In particular the boundary conditions are markedly influenced by the wettability of the substrate [11–19]. This encourages measurements on the flow dynamics through porous Vycor® with a modified surface chemistry. According to the procedure explained in section 3.1 the high surface energy of silica can easily be reduced through silanization. The critical surface tension γ_C of the treated substrate can thereby reach values down to $\sim 20\,\frac{mN}{m}$ [38]. For the untreated glass it is $\gamma_C \approx 150\,\frac{mN}{m}$. Compliant with the Zisman criterion one might predict that the dynamics of water ($\sigma \approx 70\,\frac{mN}{m}$) must be altered markedly whereas the flow of n-hexane ($\sigma \approx 20\,\frac{mN}{m}$, see Tab. A.4) should only be affected moderately.

10.2.1. Preparation

Prior to silanization the samples were flushed with trichloromethane ($CHCl_3$) several times. In the subsequent step they were exposed to a 1:9 mixture of dimethyldichlorosilane ($Si(CH_3)_2Cl_2$) and trichloromethane for about two hours. In presence of dimethyldichlorosilane low-energy methyl (CH_3) groups were substituted for the polar and consequently high-energy hydroxyl (OH) groups at the glass surface. Afterwards the samples were again flushed with trichloromethane and methanol several times.

It is important to perform this last step thoroughly since any remainder of dimethyldichlorosilane in the sample potentially reverses the silanization reaction in the presence of water, e.g., from the humidity in the laboratory. In order to further minimize the risk for such a reversal reaction the samples were dried over a stream of dry nitrogen.

The samples were characterized again by means of nitrogen sorption isotherms (see Appendix B). They reveal a reduction in the mean pore radius of approximately $4\,\text{Å}$ which is consistent with the thickness of the attached methyl groups at the pore walls. The porosity is likewise reduced. The values are listed in Tab. 10.5. For convenience I will denote the silanized samples sV5 and sV10, respectively, from now on.

Table 10.5.: Properties of the silanized Vycor® batches as extracted from measurements presented in Appendix B.

sample batch	mean pore radius r_0	volume porosity ϕ_0
sV5	$(3.0 \pm 0.1)\,\text{nm}$	0.235 ± 0.02
sV10	$(4.5 \pm 0.1)\,\text{nm}$	0.265 ± 0.02

10.2.2. Results

The results from the measurements on silanized Vycor® compared to the values from the respective untreated sample are shown in Tab. 10.6 in terms of slip lengths. The value for water in sV5 is not available since even for the highest pressures applied (70 bar) no flow through the sample could be detected. Admittedly, the dynamics of water in sV10 were still measurable, but extremely diminished. Contrary to this collapse in the dynamics of water the flow of n-hexane seems to have been even enhanced.

10.2.3. Discussion

It is obvious that the modified surface chemistry of the porous Vycor® samples significantly influences the dynamics of both liquids. Some basic discoveries are in high accordance with an NMR study of water and several alcohols in similarly treated Vycor® glass [149]. The inability of water to penetrate the sV5 sample must be traced back to the modified wettability of the substrate. Spontaneous imbibition could be observed for neither sV5 nor sV10. In consequence, a capillary depression caused by a contact angle $\theta_0 > 90°$ is substituted for the capillary rise mechanism. One can estimate a lower bound for θ_0 from the finding that even pressures up to 70 bar cannot overcome the counteracting Laplace pressure:

$$\cos\theta_0 < -\frac{\Delta p \, r_0}{2\sigma} , \qquad (10.1)$$

thus it is $\theta_0 > 98°$. Depending on the actual methyl density of the silanized surface, water can have contact angles up to 120° corresponding to a Laplace pressure of ~ 240 bar. Therefore, the complete blocking of water penetration of the sV5 sample is not surprising at all [150]. It is rather a preeminent elucidation of the magnitude of surface forces.

From this point of view the still observable flow of water through sV10 is all the more astonishing. It is possibly a mere artefact of incomplete silanization, which enables one to displace all counteracting menisci in the sample with pressures

Table 10.6.: Slip lengths b (in Å) of water and n-hexane flowing through untreated and surface silanized Vycor®, respectively. The liquids were pressurized with helium.

system	temperature	untreated	silanized
water in (s)V10	25 °C	-0.1 ± 2.2	-12.7 ± 4.8
water in (s)V5	25 °C	-2.6 ± 1.6	n/a
hexane in (s)V5	25 °C	-3.8 ± 1.8	0.3 ± 2.7
	40 °C	-4.5 ± 2.2	0.4 ± 2.8
	50 °C	-3.5 ± 2.4	0.4 ± 3.8

of maximum 70 bar. Once the sample is completely filled, water flow is principally possible. Nevertheless, the boundary conditions are significantly altered to a sticking layer of more than 1 nm thickness. This result is contrary to the presumption of slippage of water over hydrophobic surfaces [91].

One might make the supposition that thanks to the repulsive interaction between liquid and substrate a gas film whose thickness significantly exceeds the film thickness on a hydrophilic surface is established at the pore walls. This conjecture could be seen in analogy to recently found evidence of the so-called hydrophobic gap [151], that is a gas layer of molecular thickness (2 Å - 6 Å) separating water from a hydrophobic substrate even in the steady state. Hence, water flow could only occur in a core cylinder of the pore and the reduction in the volume flow rate due to the diminishment in the pore radius could surpass its increase due to potential slippage over the gas film.

The results on the flow of n-hexane in sV5 (the same sample that was used in the water experiment) are more intuitive. The reduction of the surface energy of Vycor® due to silanization weakens the attractive interaction between the surface and the alkane. This is expressed by the distinct disappearance of the sticking layer in favor of a classical no-slip boundary condition although, according to the Zisman criterion, the liquid should still totally wet the surface. Presumably a further reduction of the surface energy would finally lead to slippage. The present study however does not have the ability to prove this presumption.

10.2.4. Conclusion

It has been verified that the wettability of the substrate deeply influences the flow dynamics and boundary conditions. The observed effects range from increasing slip lengths to complete blocking of the flow. Especially the results from the alkane flow are rather promising and encourage additional studies. The application of higher alkanes, which possess higher surface tensions, would permit more detailed examinations of the influence of the wettability. Furthermore, the surface coating with fluorinated groups (instead of methyl groups) causes reductions of the critical surface tension down to $\sim 6 \frac{mN}{m}$ [38]. By these means the assessment of the conjecture of slippage over hydrophobic surfaces will be feasible. For a higher reliability of the results the density of the surface coating of each silanized samples should be examined by means of IR spectroscopy.

Part IV.

Conclusion

In this part of the thesis the results of the presented investigations will briefly be summarized. Moreover, some conceivable future studies, initiated by results gained within this thesis, will be considered.

11. CONCLUDING REMARKS

The examinations of the flow dynamics in mesopore confinement presented within this thesis revealed a wide variety of astonishing phenomenons. Most of which have conclusively been found for two different pore sizes (3.4 nm and 4.9 nm mean pore radius, respectively), thus rendering the following results particularly accurate and reliable.

To begin with, the flow of water and various linear hydrocarbons and silicon oils through silica mesopores verifies a compartmentation of the pore confined liquid: one layer of (flat lying) molecules is pinned to the pore walls whereas the residual inner region obeys classical hydrodynamics. The sticking layer is attributed to the highly attractive interaction between liquid and substrate. The molecular shape may influence the overall flow behavior via the thickness of the sticking layer, or equivalently, the molecular diameter. Over a wide range the chain length shows no impact on the dynamics, though. Reasonable future studies might address the flow behavior of 1-alcohols. Because of the additional hydroxyl group the liquid's interaction with the substrate is significantly altered and modified boundary conditions might arise therefrom.

Hints for deviations from the sticking layer toward a slip boundary condition were found for the longest hydrocarbon chain investigated in the smallest pores applied. This might be seen in analogy to the occurrence of slip in polymeric systems. Dissolved gases possibly also favor slip and, according to my results, this becomes more likely for poorly soluble gases. Furthermore, diminishing the liquid's wettability by reducing the substrate's surface energy leads to a comparable variation in the boundary conditions. The experimental evidence for these discoveries is unfortunately sparse but nonetheless encouraging and in accordance with results of a series of recent publications. Further, more systematic investigations regarding the influence of chain length, dissolved gases and wettability will be rather promising. In particular, comparative forced throughput measurements must be carried out with the pressurized gas being separated from the liquid, thus excluding any influence of dissolved gases in such experiments. Moreover, based on a thorough investigation of the silanization procedure and its impact on the surface energy, the influence of the wettability on the slip length should systematically be examined. Because of its easy handling and availability in different molecular weights polyethylene oxide is probably a well applicable polymer for a methodical study of the influence of chain length on the boundary conditions. Due to its non-negligible hygroscopy these experiments will have to be conducted in a glove box under protective atmosphere, though.

A more fundamental analysis of the spontaneous imbibition process in a complex pore network verified that the measured dynamics can solely be attributed to the capillarity-driven liquid invasion even for highly volatile liquids such as water – vapor invasion does not influence the mass uptake rate, provided appropriate measures are taken. The same is true for the relative humidity in the laboratory up to at least 50 %.

Detailed analyses of the scaling and the dynamics of the invasion front suggest the absence of any network effect in the porous Vycor® glass during the imbibition process, i.e. the lack of any lateral correlation. Many observed effects can directly be traced back to the sample's behavior analogous to an array of independent pores. This finding is further corroborated by means of gas permeation measurements. To date, the physical origins of this surprising behavior are a matter of conjecture. Nevertheless, one might speculate whether the small aspect ratios of the pore sections along with the monodirectional liquid flow prevent any correlation perpendicular to the main flow direction. Interestingly, network effects are very well noticeable when there is no such preferential direction in the system that 'breaks' the isotropy of the matrix, e.g., the steady-state behavior during capillary condensation. Simulations of the imbibition process in a complex pore network are a promising method in order to eventually shed light on this discovery. Future studies will have to address this approach.

The occurrence of a white front wandering from the sample's bottom to its top where it finally vanishes can be attributed to the generation of percolating clusters on adequate length scales that induce significant light scattering. The upper bound of this visible interface can be equated to the 'center of mass' location (with filling degree $f \approx 0.5$) as deduced from the integrated signal obtained in gravimetric measurements. Light scattering is observable almost up to the complete filling ($f \approx 1$).

The phase transition behavior of a hydrocarbon and a liquid crystal was found to be influenced in different ways by the mesoscale confinement. My study of n-tetracosane unambiguously revealed evidence of surface freezing occurring at the advancing menisci. This observation testifies to a remarkable robustness of this alignment transition of the surface molecules both upon mesoscale spatial confinement and upon self-propelled movement of the interface in a rather complex pore network. My study additionally suggests that mesopore confinement allows one to establish the surface frozen state over much larger T-ranges than possible on bulk surfaces. Again, it is reasonable to perform supplemental examinations of 1-alcohols, which show a surface freezing transition as well. However, contrary to the n-alkanes its surface crystalline phase is constituted of a *double* layer with the OH groups assembling in the middle. Thanks to the strong interaction of the hydroxyl groups with the surface such a crystalline structure and, hence, the surface freezing transition might be suppressed in the mesopore confinement.

For the rod-like liquid crystal 8OCB the strong first-order isotropic to nematic transition is replaced by a weak first-order or continuous paranematic to nematic transition in confinement. What is more, the smectic A phase is seemingly suppressed in favor of the nematic phase which is attributed to the tendency of molecular alignment along the pore axis. With regard to the observed strong influence of the rod-like molecular shape it is also worthwhile extending the present study to differently shaped liquid crystals, e.g., discoid or spherical ones.

Finally, the importance of a thorough understanding of the equilibrium behavior of the investigated molecular assemblies in confinement must be highlighted. The static confined phases were scrutinized employing many different methods ranging from sorption isotherm, birefringence, and calorimetry measurements to X-ray diffraction and transmission electron microscopy. Based on this detailed knowledge only, one is able to deduce complementary results on the liquid's nonequilibrium behavior, here the flow characteristics in mesopore confinement.

Appendix

A. Fluid Properties

Table A.1.: Fluid properties for relevant temperatures: *(Milli-Q) water.* The values are taken from: [a] [152], [b] [153].

temperature T	density ρ	viscosity η	surface tension σ
(°C)	(g/ml)	(mPa s)	(mN/m)
25	0.9970 [a]	0.8896 [a]	71.96 [b]
40	0.9919 [a]	0.6407 [a]	69.42 [b]
60	0.9817 [a]	0.4460 [a]	65.70 [b]

Table A.2.: Fluid properties for relevant temperatures: *liquid crystal 8OCB* (SYNTHON Chemicals, purity: $> 99.5\,\%$, mesophase behavior: crystalline $\xrightarrow{54\,°C}$ A $\xrightarrow{67\,°C}$ N $\xrightarrow{80\,°C}$ I). The values are taken from: [a] [154], [b] [126], [c] [155].

temperature T	density ρ	viscosity η	surface tension σ
(°C)	(g/ml)	(mPa s)	(mN/m)
60	1.0064 [a]		32.65 [a], [c]
65	1.0021 [a]		32.32 [a], [c]
70	0.9976 [a]	13.03 [b]	31.60 [a], [c]
75	0.9932 [a]	10.21 [b]	29.92 [a], [c]
80	0.9843 [a]	9.832 [b]	28.65 [a], [c]
85	0.9793 [a]	9.384 [b]	28.18 [a], [c]
91	0.9732 [a]	7.991 [b]	27.85 [a], [c]

Table A.3.: Fluid properties: *Dow Corning silicon oils* at $T = 25\,°C$. The values are taken from: [a] pycnometer measurement and [b] rheometer measurement, both by courtesy of Matthias Wolff (Saarland University, Saarbrücken, Germany), [c] [153].

DC...	density ρ	viscosity η	surface tension σ
	(g/ml)	(mPa s)	(mN/m)
...704	1.0772 [a]	48.99 [b]	32.85 [c]
...705	1.1033 [a]	218.8 [b]	35.24 [c]

Table A.4.: Densities ρ, viscosities η and surface tensions σ for the used *alkanes* at relevant temperatures T. The values are taken from: [a] [156], [b] [157], [c] [158], [d] [153], [e] cone-plate rheometer measurement by courtesy of Mario Beiner (Martin Luther University, Halle-Wittenberg, Germany), [f] Wilhelmy plate measurement by courtesy of Moshe Deutsch (Bar-Ilan University, Ramat-Gan, Israel). The info-column contains information on the respective supplier (AA: Alfa Aesar, AL: Aldrich, FL: Fluka, ME: Merck) and the alkane's minimum purity (in percent). [s] squalane (2,6,10,15,19,23-hexamethyltetracosane).

alkane	info	T (°C)	ρ (g/ml)	η (mPa s)	σ (mN/m)
n-C_6H_{14}	FL99	25	0.6532 [a]	0.2985 [a]	17.82 [a]
		40		0.2619 [a]	16.29 [a]
		50		0.2411 [a]	15.26 [a]
		60		0.2223 [a]	
n-$C_{10}H_{22}$	ME99	25	0.7271 [a]	0.8835 [a]	23.49 [a]
n-$C_{12}H_{26}$	ME99	25	0.7466 [a]	1.381 [a]	25.10 [a]
n-$C_{14}H_{30}$	AA99	25	0.7595 [a]	2.086 [a]	26.21 [a]
n-$C_{16}H_{34}$	ME99	25	0.7681 [a]	3.087 [a]	26.97 [a]
n-$C_{18}H_{38}$	FL99	36	0.7711 [a]	3.407 [a]	26.88 [a]
n-$C_{20}H_{42}$	AL99	41	0.7748 [b]	4.06 [c]	27.12 [d]
n-$C_{24}H_{50}$	FL99	52	0.7781 [b]	5.176 [e]	27.14 [f]
		53	0.7774 [b]	5.061 [e]	28.43 [f]
		54	0.7768 [b]	4.948 [e]	29.78 [f]
		57	0.7749 [b]	4.611 [e]	29.47 [f]
		59	0.7737 [b]	4.390 [e]	29.27 [f]
		61	0.7725 [b]	4.171 [e]	29.06 [f]
$C_{30}H_{62}$ [s]	ME99	30	0.8016 [b]	22.13 [c]	28.14 [d]
n-$C_{30}H_{62}$	FL98	74	0.7765 [b]	5.47 [c]	26.56 [d]
n-$C_{40}H_{82}$	AA97	91	0.7748 [b]	7.20 [c]	26.04 [d]
n-$C_{60}H_{122}$	FL98	107	0.7720 [b]	13.14 [c]	24.65 [d]

B. Matrix Characterization

The results presented within this thesis are based on accurate information on the used porous matrices. The applied samples were thoroughly characterized by means of three fundamental techniques: sorption isotherm experiments, transmission electron microscopy, and gas permeation measurments. These methods will briefly be introduced in the following.

B.1. Sorption Isotherm Measurements

A sorption isotherm is a measure of the mass uptake in a solid as the pressure of the ambient vapor is varied. It can be measured volumetrically. A ballast tank of known volume is evacuated and then connected to the temperature-controlled sample container. The pressure of the vapor in the ballast tank is measured both before and after it equilibrates with the sample, enabling both the change in volume of adsorbed gas and the ambient pressure to be determined.

In Fig. B.1 a nitrogen sorption isotherm performed at $T = 77\,\mathrm{K}$ on porous silicon is

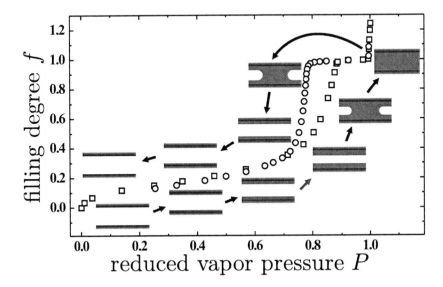

Figure B.1.: Nitrogen adsorption ($\square$) and desorption ($\bigcirc$) isotherm on porous silicon at $T = 77\,\mathrm{K} = -196\,°\mathrm{C}$. The insets illustrate the corresponding behavior of the condensate in a cylindrical pore. The red arrow indicates the non-equilibrium step of the buildup of metastable layers at the pore walls.

shown exemplarily. I plotted the filling fraction f, that is the number of nitrogen molecules adsorbed by the matrix normalized to the nitrogen amount necessary for its complete filling, versus the reduced vapor pressure $P \equiv \frac{p_e}{p_0}$. The pressure $p_0 \approx 1\,\text{bar}$ refers to the bulk vapor pressure of nitrogen at $T = 77\,\text{K}$ and p_e to the equilibration pressure after each adsorption or desorption step, respectively.

The sorption characteristics at low P are similar to those obtained on planar substrates. At higher pressures a transition to complete pore filling, which is due to capillary condensation, is observed. The transition from the adsorbate state with a thin adsorbate layer covering the inner walls of the pores and vapor in the central part of the pores to the capillary condensed state is of first order and, hence, shows an intrinsic hysteresis on adsorption and desorption [159]. This distinctive hysteresis is characteristic of this class of porous material [160].

The fundamentals of the capillary condensation transition were worked out by Will Saam and Milton Cole [161]. In contrast to adsorption on a planar substrate two essential points have to be considered for adsorption on the inner walls of cylindrical tubes: (i) The area of the adsorbate-vapor interface decreases with increasing thickness of the adsorbed layer; thereby, the adsorbate-vapor surface tension enters into the problem. (ii) The adsorbate-substrate interaction po-

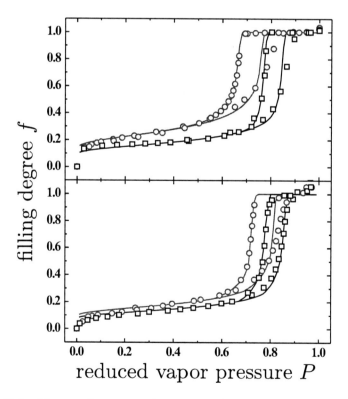

Figure B.2.: Nitrogen (upper panel) and water (lower panel) sorption isotherms on V10 ($\square$) and V5 ($\bigcirc$) carried out at $T = -196\,^{\circ}\text{C}$ and $T = 4\,^{\circ}\text{C}$, respectively. Solid lines are fits according to a mean field model (Saam-Cole).

tential is different from the planar substrate. In particular, the gradient of the potential vanishes in the center of the pore. The transition from the film state to the capillary condensed state occurs when the cost of adsorption energy – because molecules move farther away from the substrate – equals the gain of interfacial energy related to the reduction of the vapor-condensate interface due to the formation of a concave meniscus. Thus, whenever the adsorbate layer grows to a thickness close to the pore radius a phase transition will necessarily take place between the condensed film state and the filled pore state [159]. The occurrence of the hysteresis is caused by the buildup of metastable states during vapor adsorption. The transition to the capillary condensate with concave menisci is delayed in favor of the growth of additional, but thermodynamically not stable layers.

The radius r of the concave menisci is closely connected to the corresponding reduced vapor pressure via the Kelvin equation

$$\ln\left(\frac{p}{p_0}\right) = \ln P = -\frac{2\sigma V_{\mathrm{mol}}}{RTr} \tag{B.1}$$

with the surface tension σ, the molar volume V_{mol} of the condensate and the gas constant R. This relation states that smaller pores are characterized by lower reduced vapor pressures P. This behavior is illustrated by the isotherms on V5 and V10 shown in Fig. B.2. For both adsorbents (nitrogen and water) the precipitous drop in desorption is distinctly shifted towards lower P for V5 as compared with V10. Though, the absolut extent of the shift sensitively depends on the respective adsorbent via the liquid's surface tension, its molar volume and the measuring temperature T.

The Kelvin equation principally permits one to determine a pore size distribution from the desorption branch. The respective change in the filling degree f for a given reduced pressure P can be associated with a probability of the corresponding radius r [162]. This radius however is smaller than the real pore radius since the menisci always coexist with a certain wall coating (see insets in Fig. B.1) whose thickness cannot be known beforehand.

A more reliable analysis of the conducted isotherms can be accomplished through an evaluation within the mean field model proposed by Will Saam and Milton Cole [161]. Compliant with a trial-and-error technique I calculated isotherms corresponding to preset pore size distributions $P(r)$ and compared the results to the measurements. In several loops the initial distributions were adjusted until calculations and experiments adequately coincided (for more details on the exact procedure see Refs. [163, 164]). The pore size distributions of V5 and V10 already presented in Fig. 3.2 are based on the nitrogen sorption isotherms shown in Fig. B.2. The corresponding Saam-Cole predictions are indicated by the solid lines.

The impact of the silanization process is preeminently elucidated by the nitrogen sorption isotherms shown in Fig. B.3. The reduction in the pore radius due to the additionally attached methyl groups is distinctly displayed by the shift of the inset of capillary condensation towards lower P. For both V5 and V10 a

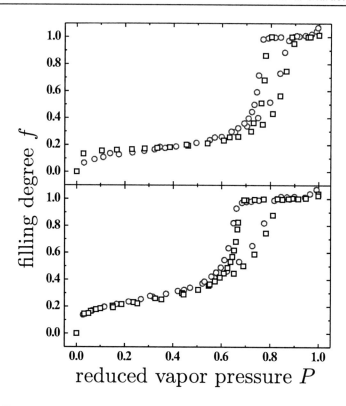

Figure B.3.: Nitrogen sorption isotherms at $T = -196\,^{\circ}\mathrm{C}$ on untreated ($\square$) and silanized ($\bigcirc$) V10 (upper panel) and V5 (lower panel), respectively.

reduction of the pore size of $\sim 4\,\text{Å}$ is observed. The overall uptake of adsorbent and, accordingly, the sample's porosity ϕ_0 is similarly diminished.

B.2. Transmission Electron Microscopy

Transmission electron micrographs from porous silicon membranes were taken in the group of Rainer Birringer[1]. Prior to examination the samples were argon ion milled to a maximum thickness of $\sim 50\,\text{nm}$ in order to ensure a sufficient transmission rate of the electrons. The image contrast is due to absorption of electrons in the material, and consequently due to the thickness and composition of the material. Pores therefore correspond to the bright areas in the micrographs (Fig. 3.6).

The diffraction pattern in Fig. B.4 elucidates the crystalline structure of the remaining silicon skeleton of the membrane. The crystalline silicon walls call for irregular channel perimeters [43] (see Fig. 3.6). Anyhow, as a first approximation the nanochannel's cross section can be described as circular with a diameter of $\sim 12\,\text{nm}$. This is in accordance with the analysis of the nitrogen sorption

---

[1]Labor für analytische Elektronenmikroskopie at Saarland University, Saarbrücken, Germany.

Figure B.4.: Diffraction pattern of porous silicon taken in the TEM setup. Courtesy of Jörg Schmauch, Saarland University, Saarbrücken, Germany.

experiments just presented.

B.3. Gas Permeation Measurements

Measurements on the gas flow through a mesoporous membrane allow for the determination of the diffusion constant D of the respective gas. Of course, the flow characteristics of a gas on the one hand or of a liquid on the other hand are similarly influenced by the membrane's tortuosity τ. This can be expressed in terms of the measured diffusion constant D compared with $D_{\tau=1}$, the diffusion constant of a hypothetical sample with aligned and straight pores [1]:

$$\tau \equiv \frac{D_{\tau=1}}{D} \, . \tag{B.2}$$

Based on this relation gas permeation measurements render possible the determination of the sample's tortuosity.

The experimental setup of the gas flow apparatus depicted in Fig. B.5 consists of a copper cell with an inlet and outlet opening. Inlet and outlet are connected via stainless steel capillaries with two gas reservoirs, R1 and R2, of an all-metal gas handling. Two pneumatic valves V1 and V2 are used to open and close the connections between the sample cell and R1 and the sample cell and R2. Four temperature-controlled capacitive pressure gauges enable one to measure the gas pressures in R1 and R2, p_1 and p_2, respectively, over a wide pressure range ($5\,\mu$bar $< p < 1\,$bar) with an accuracy of $1\,\mu$bar. The cell is mounted in a closed-cycle helium cryostat in order to control the temperature between $10\,$K and $300\,$K with an accuracy of $1\,$mK.

The gas relaxation between R1 (with starting pressure $p_1(t=0\,\mathrm{s}) \equiv p_\mathrm{s}$) and R2 ($p_2(t=0\,\mathrm{s}) = 0\,$bar) *without* a mesoporous membrane in the copper cell is plotted in Fig. B.6 (a) for two different starting pressures. One can define a relaxation time τ_r (that must not be confused with the tortuosity τ) that characterizes

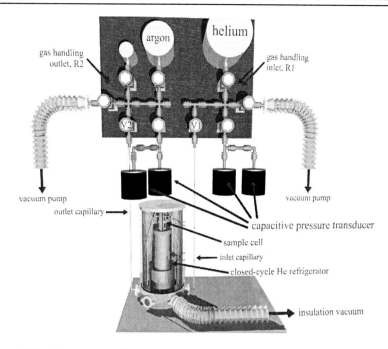

Figure B.5.: Raytracing illustration of the gas permeation setup. For more details see Refs. [2, 165]

each measurement via the recipe $p_1(t = \tau_r) - p_2(t = \tau_r) \equiv \frac{1}{10} p_s$. The so obtained relaxation times plotted in Fig. B.6 (c) show a nontrivial behavior, which can be understood after sorting the results in terms of the dimensionless Knudsen number Kn. The latter quantifies the degree of rarefaction of the gas by comparing its mean free path $\lambda(p, T)$ at a given pressure and temperature with the characteristic length scale of the flow geometry $\mathcal{L}$: $\mathrm{Kn} \equiv \frac{\lambda}{\mathcal{L}}$.

For $\mathrm{Kn} \ll 1$ the mean free path is much smaller than the dimensions of the flow path. Just as in a liquid, collisions between molecules prevail and the dynamics can be described within the continuum mechanical theory for a compressible fluid. Conversely, for $\mathrm{Kn} \gg 1$ collisions of the molecules with the confining walls dominate and the dynamics turn to a ballistic movement of the molecules also referred to as Knudsen diffusion [166, 167]. The intermediate regime ($\mathrm{Kn} \approx 1$) is not fully understood yet. Nonetheless, as it turns out, a model proposed by George Karniadakis and Ali Beskok [168] permits one to preeminently describe the transition from continuum to free molecular flow [43] (see solid lines in Fig. B.6 (c), which correspond to calculations based on this model).

For the pure flow of the gas through the supplying capillaries ($\mathcal{L} = \mathcal{O}(1\,\mathrm{mm})$) and for the pressures probed in this study the resultant Knudsen numbers $\mathrm{Kn_c}$ pass through all regimes outlined above (see bottom axis in Fig. B.6 (c)). Thus, solely the capillaries are responsible for the particular behavior of the relaxation times τ_r. This is elucidated by the comparison with the relaxation behavior after assembling the mesoporous membrane in the copper cell. Two measurements

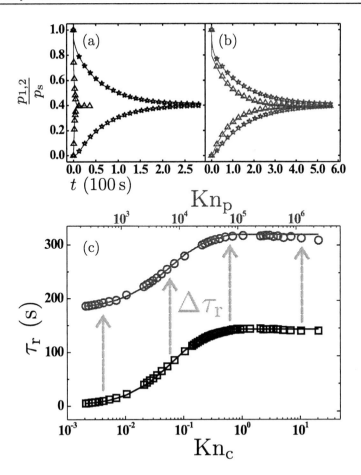

Figure B.6.: Normalized helium pressure relaxations in R1 and R2 at $T = 297\,\mathrm{K}$ for starting pressures $p_\mathrm{s} = 0.1\,\mathrm{mbar}$ ($\star$ and $\star$) and $p_\mathrm{s} = 100\,\mathrm{mbar}$ ($\triangle$ and $\triangle$) without (a) and with (b) a mesoporous membrane in the copper cell. (c): Pressure relaxation time τ_r as determined from measurements without ($\square$) and with mesoporous membrane ($\bigcirc$) versus Knudsen number in the capillaries $\mathrm{Kn_c}$ (bottom axis) and in the mesopores $\mathrm{Kn_p}$ (top axis). The lines in (a), (b), and (c) represent calculated $p(t)$ and $\tau_\mathrm{r}(\mathrm{Kn})$ values.

are exemplarily shown in Fig. B.6 (b) and the corresponding relaxation times are additionally plotted in Fig. B.6 (c).

It is obvious that the only effect is a nearly constant shift of τ_r towards higher values. This is not surprising at all. Because of the significantly smaller $\mathcal{L} = \mathcal{O}(10\,\mathrm{nm})$ the Knudsen numbers for the flow through the mesoporous membrane $\mathrm{Kn_p}$ all hold $\mathrm{Kn_p} \gg 1$ (see top axis in Fig. B.6 (c)). The dynamics therefore are purely diffusive and from the shift $\Delta\tau_\mathrm{r}$ one can deduce the diffusion constant D of the respective gas. Using Eq. (B.2) this finally yields the sample's tortuosity τ. The measurements performed in the course of the diploma thesis of Stefan Bommer revealed $\tau \approx 3.9 \pm 0.4$ for both V5 and V10 [2].

B.4. Assessment of the Pore Connectivity in Porous Vycor®

The gas permeation measurements just presented enable one to assess the degree of pore connectivity in porous Vycor® compared to porous silicon. For this purpose I filled the sample cell at $T = 100\,\mathrm{K}$ with a well-known amount of argon gas. Afterwards the sample was slowly cooled down to $T = 40\,\mathrm{K}$, well below argon's bulk triple point at $T \approx 84\,\mathrm{K}$, in order to make sure that the complete offered amount of gas condensed in the porous substrate. The actual gas flow was measured with helium gas, which, contrary to argon, does not condense at $T = 40\,\mathrm{K}$ yet.

Applying this procedure one is able to change the pore morphology. This can be understood referring to the statements brought up in the discussion of the sorption isotherm measurements. For low filling degrees f the gas builds layers at the pore walls. Thereby the effective pore radius is reduced. At the same time the formerly rough walls might be rendered more smooth [169]. At higher filling degrees capillary condensation sets in and pores with compatible pore radii are blocked. For porous silicon this must result in an instant drop in the measured flow rate. Though, for a pore network like in porous Vycor® there might still be some alternative route that can be taken by the gas. Because of such potential detours no abrupt drop should be observed in this particular case.

The measured diffusion coefficients D normalized with the respective coefficient for $f = 0$ are plotted in Fig. B.7 for a porous Vycor® as well as for a porous silicon

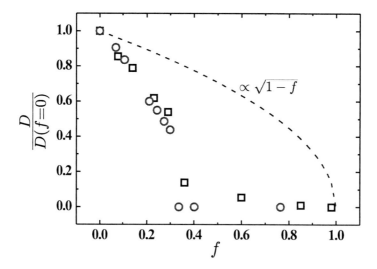

Figure B.7.: Diffusion coefficients D at $T = 40\,\mathrm{K}$ of helium flowing through V5 ($\square$) and porous silicon ($\bigcirc$) as a function of the filling degree f of argon. The coefficients are normalized with respect to $D(f = 0)$. The measurements on V5 were carried out by Stefan Bommer in the course of his diploma thesis (see [2]).

sample. Both sample types are characterized by a similarly steep diminishment in D for low f. This finding is not too surprising if one assumes this behavior to be based on a reduction in the effective pore radius due to the adsorption of gas at the pore walls. Such a coating influences sieve-like and sponge-like membranes alike.

Nevertheless, a more detailed analysis of this initial regime reveals that the observed decrease in D can not solely be attributed to a decrease in the effective pore radius r_{eff}. This can easily be verified: the effective cross-sectional area of a pore for a given filling degree f is given by $(1-f)\,\pi r_0^2 = \pi r_{\text{eff}}^2$, and hence $r_{\text{eff}} = r_0\sqrt{1-f}$. Because of $D \propto r$ [43] the normalized quantity calculated in Fig. B.7 should coincide with

$$\frac{D(f)}{D(f=0)} = \frac{D(r=r_{\text{eff}})}{D(r=r_0)} = \frac{r_{\text{eff}}}{r_0} = \sqrt{1-f}\,. \qquad \text{(B.3)}$$

This relation corresponds to the dashed line in Fig. B.7. It is evident that the actually observed diminishment is far more pronounced than expected from this simple consideration. In consequence, different mechanisms must further influence the available flow topologies – apparently for porous silicon *and* porous Vycor® in a similar way.

The inset of extensive capillary condensation is expected for $f \approx 0.35$ as extracted from argon sorption isotherms [170]. For this filling degree at the latest a sieve-like membrane should completely be blocked. Indeed, no measurable diffusion rate can be observed in porous silicon for $f > 0.35$. But what are the reasons for the marked diminishment for $f < 0.35$?

Here, one has to keep in mind the non-negligible pore size distribution of the samples and that the pore radii are not uniform along the channel's long axis. One might assume that there is a constriction in each pore that can establish plugs already for $f < 0.35$, thus blocking the whole pore. Likewise, it might be an artefact of the preparation method. Studies on the heat capacity of partial fillings of argon in a mesoporous host revealed a so-called delayering transition at $T = 66\,\text{K}$ [171, 172]. It was reported that stable layers of argon became metastable during cooling and finally, at $T = 66\,\text{K}$, transformed into stable capillary condensate. This might also have happened in my study while cooling down the sample from $T = 100\,\text{K}$ to $T = 40\,\text{K}$. It would result in a premature buildup of plugs in the pores, which might entail the observed additional drop in the diffusion coefficient.

For a sieve-like membrane (porous silicon) this argumentation is conclusive. In a complex pore network such premature pore blocking should be less effective due to possible detours that can be taken. Though, since the porous Vycor® sample behaves similarly to the porous silicon membrane, the impact of the pore connectivity can only be weakly distinctive. The observed behavior suggests that, to a considerable extent, Vycor® glass behaves analogous to an array of independent pores. This discovery reflects results from a detailed analysis of the dynamics of invading liquid interfaces presented in chapter 7.

Pore connectivity seemingly plays a subordinate role only. It can, however, not be completely neglected as even beyond the inset of extensive capillary condensation diffusion through the pore network can be observed. The residual diffusion coefficient is $\sim \frac{1}{10} D(f = 0)$ for $f = 0.35$ and even for $f \approx 0.6$ finite gas flux was measured. This result can directly be traced back to the pore connectivity in Vycor® glass.

List of Tables

List of Figures

Bibliography

[1] Lin MY, Abeles B, Huang JS, Stasiewski HE, and Zhang Q. Visocus flow and diffusion of liquids in microporous glasses. *Phys. Rev. B* **46**, 10701 (1992).

[2] Bommer S. *Gas- und Proteinpermeabilitätsmessungen an biomimetischen Festkörpermembranen.* Diploma thesis, Saarland University, Saarbrücken, Germany (2008).

[3] Crossley RA, Schwartz LM, and Banavar JR. Image-based models of porous media: Application to Vycor glass and carbonate rocks. *Appl. Phys. Lett.* **59**, 3553 (1991).

[4] Debye P and Cleland RL. Flow of liquid hydrocarbons in porous Vycor. *J. Appl. Phys.* **30** (6), 843 (1959).

[5] Israelachvili J. Measurement of the viscosity of liquids in very thin films. *J. Colloid Interface Sci.* **110**, 263 (1986).

[6] Horn RG, Smith DT, and Haller W. Surface forces and viscosity of water measured between silica sheets. *Chem. Phys. Lett.* **162**, 404 (1989).

[7] Raviv U, Laurat P, and Klein J. Fluidity of water confined to subnanometre films. *Nature* **413** (6851), 51 (2001).

[8] Gupta SA, Cochran HD, and Cummings PT. Shear behavior of squalane and tetracosane under extreme confinement. III. Effect of confinement on viscosity. *J. Chem. Phys.* **107**, 10335 (1997).

[9] Fisher LR, Gamble RA, and Middlehurst J. The Kelvin equation and the capillary condensation of water. *Nature* **290**, 575 (1981).

[10] Fradin C *et al.* Reduction in the surface energy of liquid interfaces at short length scales. *Nature* **403**, 871 (2000).

[11] Barrat JL and Bocquet L. Large slip effect at a nonwetting fluid-solid interface. *Phys. Rev. Lett.* **82**, 4671 (1999).

[12] Pit R, Hervet H, and Leger L. Direct experimental evidence of slip in hexadecane: Solid interfaces. *Phys. Rev. Lett.* **85**, 980 (2000).

[13] Cieplak M, Koplik J, and Banavar JR. Boundary conditions at a fluid-solid interface. *Phys. Rev. Lett.* **86**, 803 (2001).

[14] Tretheway DC and Meinhart CD. Apparent fluid slip at hydrophobic microchannel walls. *Phys. Fluids* **14**, L9 (2002).

[15] Cho JHJ, Law BM, and Rieutord F. Dipole-dependent slip of Newtonian liquids at smooth solid hydrophobic surfaces. *Phys. Rev. Lett.* **92**, 166102 (2004).

[16] Schmatko T, Hervet H, and Leger L. Friction and slip at simple fluid-solid interfaces: The roles of the molecular shape and the solid-liquid interaction. *Phys. Rev. Lett.* **94**, 244501 (2005).

[17] Fetzer R and Jacobs K. Slippage of Newtonian liquids: Influence on the dynamics of dewetting thin films. *Langmuir* **23**, 11617 (2007).

[18] Voronov RS, Papavassiliou DV, and Lee LL. Review of fluid slip over superhydrophobic surfaces and its dependence on the contact angle. *Ind. Eng. Chem. Res.* **47**, 2455 (2008).

[19] Maali A, Cohen-Bouhacina T, and Kellay H. Measurement of the slip length of water flow on graphite surface. *Appl. Phys. Lett.* **92**, 053101 (2008).

[20] Zhu Y and Granick S. Rate-dependent slip of Newtonian liquid at smooth surfaces. *Phys. Rev. Lett.* **87**, 096105 (2001).

[21] Craig VSJ, Neto C, and Williams DRM. Shear-dependent boundary slip in an aqueous Newtonian liquid. *Phys. Rev. Lett.* **87**, 054504 (2001).

[22] Priezjev NV and Troian SM. Molecular origin and dynamic behavior of slip in sheared polymer films. *Phys. Rev. Lett.* **92**, 018302 (2004).

[23] Priezjev NV. Rate-dependent slip boundary conditions for simple fluids. *Phys. Rev. E* **75**, 051605 (2007).

[24] Vinogradova OI and Yakubov GE. Surface roughness and hydrodynamic boundary conditions. *Phys. Rev. E* **73**, 045302(R) (2006).

[25] Zhu Y and Granick S. Limits of the hydrodynamic no-slip boundary condition. *Phys. Rev. Lett.* **88**, 106102 (2002).

[26] Bonaccurso E, Butt HJ, and Craig VSJ. Surface roughness and hydrodynamic boundary slip of a Newtonian fluid in a completely wetting system. *Phys. Rev. Lett.* **90**, 144501 (2003).

[27] Granick S, Zhu YX, and Lee H. Slippery questions about complex fluids flowing past solids. *Nat. Mater.* **2**, 221 (2003).

[28] Dammer SM and Lohse D. Gas enrichment at liquid-wall interfaces. *Phys. Rev. Lett.* **96**, 206101 (2006).

[29] Cheikh C and Koper G. Stick-slip transition at the nanometer scale. *Phys. Rev. Lett.* **91**, 156102 (2003).

[30] Lauga E, Brenner MP, and Stone HA. *Handbook of Experimental Fluid Dynamics*, chapter 15 Microfluidics: The No-Slip Boundary Condition (Springer, New York, 2006).

[31] Neto C, Evans DR, Bonaccurso E, Butt HJ, and Craig VSJ. Boundary slip in Newtonian liquids: A review of experimental studies. *Rep. Prog. Phys.* **68**, 2859 (2005).

[32] Todd BD and Evans DJ. The heat-flux vector for highly inhomogeneous nonequilibrium fluids in very narrow pores. *J. Chem. Phys.* **103** (22), 9804 (1995).

[33] Travis KP, Todd BD, and Evans DJ. Departure from Navier-Stokes hydrodynamics in confined liquids. *Phys. Rev. E* **55** (4), 4288 (1997).

[34] Dimitrov DI, Milchev A, and Binder K. Capillary rise in nanopores: Molecular dynamics evidence for the Lucas-Washburn equation. *Phys. Rev. Lett.* **99**, 054501 (2007).

[35] Elmer TH. *Porous and reconstructed glasses*, volume 4 of *Engineered Materials Handbook*, page 427 (ASM International, Materials Park, OH, 1992).

[36] Ito Y, Yamashina T, and Nagasaka M. Structure of porous Vycor glass and its adsorption characteristics of water – An application of positron annihilation method. *Appl. Phys.* **6**, 323 (1975).

[37] Levitz P, Ehret G, Sinha SK, and Drake JM. Porous Vycor glass: The microstructure as probed by electron microscopy, direct energy transfer, small-angle scattering, and molecular adsorption. *J. Chem. Phys.* **95**, 6151 (1991).

[38] de Gennes PG, Brochard-Wyart F, and Quere D. *Capillarity and Wtting Phenomena: Drops, Bubbles, Pearls, Waves* (Springer, New York, 2004).

[39] Lehmann V and Gösele U. Porous silicon formation: A quantum wire effect. *Appl. Phys. Lett.* **58**, 856 (1991).

[40] Lehmann V, Stengl R, and Luigart A. On the morphology and the electrochemical formation mechanism of mesoporous silicon. *Mater. Sci. Eng. B* **69-70**, 11 (2000).

[41] Kityk AV *et al.* Continuous paranematic-to-nematic ordering transitions of liquid crystals in tubular silica nanochannels. *Phys. Rev. Lett.* **101**, 187801 (2008).

[42] Henschel A, Hofmann T, Huber P, and Knorr K. Preferred orientations and stability of medium length n-alkanes solidified in mesoporous silicon. *Phys. Rev. E* **75**, 021607 (2007).

[43] Gruener S and Huber P. Knudsen diffusion in silicon nanochannels. *Phys. Rev. Lett.* **100**, 064502 (2008).

[44] Kumar P *et al.* Tuning the pore wall morphology of mesoporous silicon. *J. Appl. Phys.* **103**, 024303 (2008).

[45] Koch GW, Sillett SC, Jennings GM, and Davis SD. The limits to tree height. *Nature* **428**, 851 (2004).

[46] Wallacher D. *FT-Infrarot-Untersuchungen an Stickstoff absorbiert in nanoporösen Gläsern*. Diploma thesis, Saarland University, Saarbrücken, Germany (1997).

[47] Pooley CM, Kusumaatmaja H, and Yeomans JM. Modelling capillary filling dynamics using lattice Boltzmann simulations. *Eur. Phys. J. Special Topics* **171**, 63 (2009).

[48] Gruener S, Hofmann T, Wallacher D, Kityk AV, and Huber P. Capillary rise of water in hydrophilic nanopores. *Phys. Rev. E* **79**, 067301 (2009).

[49] Lucas R. Über das Zeitgesetz des kapillaren Aufstiegs von Flüssigkeiten. *Kolloid Zeitschrift* **23**, 15 (1918).

[50] Washburn EW. The dynamics of capillary flow. *Phys. Rev.* **17**, 273 (1921).

[51] Bell JM and Cameron F. The flow of liquids through capillary spaces. *J. Phys. Chem.* **10**, 658 (1906).

[52] Reyssat M, Courbin L, Reyssat E, and Stone HA. Imbibition in geometries with axial variations. *J. Fluid Mech.* **615**, 335 (2008).

[53] Wiltzius P, Bates FS, Dierker SB, and Wignall GD. Structure of porous Vycor glass. *Phys. Rev. A* **36**, 2991 (1987).

[54] Page JH, Liu J, Abeles B, Deckman HW, and Weitz DA. Pore-space correlations in capillary condensation in Vycor. *Phys. Rev. Lett.* **71**, 1216 (1993).

[55] Soprunyuk VP, Wallacher D, Huber P, Knorr K, and Kityk AV. Freezing and melting of Ar in mesopores studied by optical transmission. *Phys. Rev. B* **67**, 144105 (2003).

[56] Soprunyuk VP *et al.* Optical transmission measurements on phase transitions of O_2 and CO in mesoporous glass. *J. Low Temp. Phys.* **134**, 1043 (2004).

[57] Gelb LD and Hopkins AC. Dynamics of the capillary rise in nanocylinders. *Nano Lett.* **2** (11), 1281 (2002).

[58] Supple S and Quirke N. Rapid imbibition of fluids in carbon nanotubes. *Phys. Rev. Lett.* **90** (21), 214501 (2003).

[59] Supple S and Quirke N. Molecular dynamics of transient oil flows in nanopores i: Imbibition speeds for single wall carbon nanotubes. *J. Chem. Phys.* **17**, 8571 (2004).

[60] Whitby M and Quirke N. Fluid flow in carbon nanotubes and nanopipes. *Nat. Nanotechnol.* **2**, 87 (2007).

[61] Majumder M, Chopra N, Andrews R, and Hinds BJ. Enhanced flow in carbon nanotubes. *Nature* **438**, 44 (2005).

[62] Holt JK *et al.* Fast mass transport through sub-2-nanometer carbon nanotubes. *Science* **312**, 1034 (2006).

[63] Ralston J, Popescu M, and Sedev R. Dynamics of wetting from an experimental point of view. *Annu. Rev. Mater. Res.* **38**, 23 (2008).

[64] Sikalo S, Wilhelm HD, Roisman IV, Jakirlic S, and Tropea C. Dynamic contact angle of spreading droplets: Experiments and simulations. *Phys. Fluids* **17**, 062103 (2005).

[65] Latva-Kokko M and Rothman DH. Scaling of dynamic contact angles in a lattice-Boltzmann model. *Phys. Rev. Lett.* **98**, 254503 (2007).

[66] Wang XD, Peng XF, and Wang BX. Effect of solid surface properties on dynamic contact angles. *Heat Trans. Asian Res.* **35**, 1 (2006).

[67] Hofmann RL. A study of the advancing interface. I. Interface shape in liquid-gas systems. *J. Colloid Interface Sci.* **50**, 228 (1975).

[68] Jiang TS, Oh SG, and Slattery JC. Correlation for dynamic contact angle. *J. Colloid Interface Sci.* **69**, 74 (1979).

[69] Zhmud BV, Tibert F, and Hallstensson K. Dynamics of capillary rise. *J. Colloid Interface Sci.* **228**, 263 (2000).

[70] Martic G *et al.* A molecular dynamics simulation of capillary imbibition. *Langmuir* **18**, 7971 (2002).

[71] Kusumaatmaja H, Pooley CM, Girado S, Pisignano D, and Yeomans JM. Capillary filling in patterned channels. *Phys. Rev. E* **77**, 067301 (2008).

[72] http://einrichtungen.physik.tu-muenchen.de/antares/ .

[73] Drake JM and Klafter J. Dynamics of confined molecular systems. *Phys. Today* **43**, 46 (1990).

[74] Bellissent-Funel MC. Status of experiments probing the dynamics of water in confinement. *Eur. Phys. J. E* **12**, 83 (2003).

[75] Alava M, Dube M, and Rost M. Imbibition in disordered media. *Adv. Phys.* **53**, 83 (2004).

[76] Brovchenko I and Oleinikova A. *Interfacial and Confined Water* (Elsevier, Amsterdam, 2008).

[77] Ball P. Water as an active constituent in cell biology. *Chem. Rev.* **108**, 74 (2008).

[78] Wheeler TD and Stroock AD. The transpiration of water at negative pressures in a synthetic tree. *Nature* **455**, 208 (2008).

[79] van Delft KM *et al.* Micromachined Fabry-Perot interferometer with embedded nanochannels for nanoscale fluid dynamics. *Nano Lett.* **7**, 345 (2007).

[80] Cupelli C *et al.* Dynamic capillary wetting studied with dissipative particle dynamics. *New J. Phys.* **10**, 043009 (2008).

[81] Chibbaro S. Capillary filling with pseudo-potential binary Lattice-Boltzmann model. *Eur. Phys. J. E* **27**, 99 (2008).

[82] Heinbuch U and Fischer J. Liquid flow in pores: Slip, no-slip, or multilayer sticking. *Phys. Rev. A* **40**, 1144 (1989).

[83] Ricci MA, Bruni F, Gallo P, Rovere M, and Soper AK. Water in confined geometries: Experiments and simulations. *J. Phys.: Condens. Matter* **12**, A345 (2000).

[84] Gallo P, Rovere M, and Spohr E. Supercooled confined water and the mode coupling crossover temperature. *Phys. Rev. Lett.* **85**, 4317 (2000).

[85] Vichit-Vadakan W and Scherer GW. Measuring permeability of rigid materials by a beam-bending method: II, Porous glass. *J. Am. Ceram. Soc.* **83**, 2240 (2000).

[86] Gallo P, Ricci MA, and Rovere M. Layer analysis of the structure of water confined in Vycor glass. *J. Chem. Phys.* **116**, 342 (2002).

[87] Castrillon SRV, Giovambattista N, Aksay IA, and Debenedetti PG. Effect of surface polarity on the structure and dynamics of water in nanoscale confinement. *J. Phys. Chem. B* **113**, 1438 (2009).

[88] Fouzri A, Dorbez-Sridi R, and Oumezzine M. Water confined in silica gel and in Vycor glass at low and room temperature, X-ray diffraction study. *J. Chem. Phys.* **116**, 791 (2002).

[89] Li TD, Gao J, Szoszkiewicz R, Landman U, and Riedo E. Structured and viscous water in subnanometer gaps. *Phys. Rev. B* **75**, 115415 (2007).

[90] Lasne D *et al.* Velocity profiles of water flowing past solid glass surfaces using fluorescent nanoparticles and molecules as velocity probes. *Phys. Rev. Lett.* **100**, 214502 (2008).

[91] Vinogradova OI. Slippage of water over hydrophobic surfaces. *Int. J. Miner. Process.* **56**, 31 (1999).

[92] Tombari E, Salvetti G, Ferrari C, and Johari GP. Thermodynamic functions of water and ice confined to 2 nm radius pores. *J. Chem. Phys.* **122**, 104712 (2005).

[93] Israelachvili J. *Intermolecular & Surface Forces* (Academic Press, London, 2006).

[94] Stanley HE *et al.* Statistical physics and liquid water at negative pressures. *Physica A* **315**, 281 (2002).

[95] Tanaka H. Thermodynamic anomaly and polyamorphism of water. *Europhys. Lett.* **50**, 340 (2000).

[96] Shin K *et al.* Enhanced mobility of confined polymers. *Nat. Mater.* **6**, 961 (2007).

[97] Dorris GM and Gray DG. Adsorption of hydrocarbons on silica-supported water surfaces. *J. Phys. Chem.* **85**, 3628 (1981).

[98] Chan DYC and Horn RG. The drainage of thin liquid-films between solid-surfaces. *J. Chem. Phys.* **83**, 5311 (1985).

[99] Christenson HK, Horn RG, and Israelachvili JN. Measurement of forces due to structure in hydrocarbon liquids. *J. Colloid Interface Sci.* **88**, 79 (1982).

[100] Georges JM, Millot S, Loubet JL, and Tonck A. Drainage of thin liquid-films between relatively smooth surfaces. *J. Chem. Phys.* **98**, 7345 (1993).

[101] Cui ST, Cummings PT, and Cochran HD. Molecular simulation of the transition from liquidlike to solidlike behavior in complex fluids confined to nanoscale gaps. *J. Chem. Phys.* **114**, 7189 (2001).

[102] Mo HD, Evmenenko G, and Dutta P. Ordering of liquid squalane near a solid surface. *Chem. Phys. Lett.* **415** (1-3), 106 (2005).

[103] Basu S and Satija SK. In-situ X-ray reflectivity study of alkane films grown from the vapor phase. *Langmuir* **23**, 8331 (2007).

[104] del Campo V *et al.* Structure and growth of vapor-deposited n-dotriacontane films studied by X-ray reflectivity. *Langmuir* **25**, 12962 (2009).

[105] Stevens MJ *et al.* Comparison of shear flow of hexadecane in a confined geometry and in bulk. *J. Chem. Phys.* **106** (17), 7303 (1997).

[106] Lenormand R. Liquids in porous media. *J. Phys.: Condens. Matter* **2**, SA79 (1990).

[107] Barabasi AL and Stanley HE. *Fractal Concepts in Surface Growth* (Cambridge University Press, New York, 1995).

[108] Dube M *et al.* Conserved dynamics and interface roughening in spontaneous imbibition: A phase field model. *Eur. Phys. J. B* **15**, 701 (2000).

[109] Ocko BM *et al.* Surface freezing in chain molecules: Normal alkanes. *Phys. Rev. E* **55**, 3164 (1997).

[110] Earnshaw JC and Hughes CJ. Surface-induced phase-transition in normal alkane fluids. *Phys. Rev. A* **46**, R4494 (1992).

[111] Sirota EB, Wu XZ, Ocko BM, and Deutsch M. What drives the surface freezing in alkanes? *Phys. Rev. Lett.* **79** (3), 531 (1997).

[112] Wu XZ *et al.* Surface tension measurements of surface freezing in liquid normal alkanes. *Science* **261**, 1018 (1993).

[113] Wu XZ, Sirota EB, Sinha SK, Ocko BM, and Deutsch M. Surface crystallization of liquid normal alkanes. *Phys. Rev. Lett.* **70**, 958 (1993).

[114] Bowick MJ, Nelson DR, and Travesset A. Interacting topological defects on frozen topographies. *Phys. Rev. B* **62**, 8738 (2000).

[115] Prasad S and Dhinojwala A. Rupture of a two-dimensional alkane crystal. *Phys. Rev. Lett.* **95** (11), 117801 (2005).

[116] Waheed M, Lavine MS, and Rutledge GC. Molecular simulation of crystal growth in n-eicosane. *J. Chem. Phys.* **116**, 2301 (2002).

[117] Alba-Simionesco C *et al.* Effects of confinement on freezing and melting. *J. Phys.: Condens. Matter* **18**, R15 (2006).

[118] Huber P, Wallacher D, Albers J, and Knorr K. Quenching of lamellar ordering in an n-alkane embedded in nanopores. *Europhys. Lett.* **65**, 351 (2004).

[119] Huber P, Soprunyuk VP, and Knorr K. Structural transformations of even-numbered n-alkanes confined in mesopores. *Phys. Rev. E* **74**, 031610 (2006).

[120] Hughes CJ and Earnshaw JC. Light-scattering study of a surface-induced phase-transition in alkane fluids. *Phys. Rev. E* **47**, 3485 (1993).

[121] Huber P *et al.* Faraday instability in a surface-frozen liquid. *Phys. Rev. Lett.* **94**, 184504 (2005).

[122] Labajos-Broncano L, Antequera-Barroso JA, Gonzalez-Martin ML, and Bruque JM. An experimental study about the imbibition of aqueous solutions of low concentration of a non-adsorbable surfactant in a hydrophilic porous medium. *J. Colloid Interface Sci.* **301**, 323 (2006).

[123] Lee KS, Ivanova N, Starov VM, Hilal N, and Dutschk V. Kinetics of wetting and spreading by aqueous surfactant solutions. *Adv. Colloid Interface Sci.* **144**, 54 (2008).

[124] Miesowicz M. The 3 coefficients of viscosity of anisotropic liquids. *Nature* **158**, 261 (1946).

[125] Graf HH, Kneppe H, and Schneider F. Shear and rotational viscosity coefficients of two nematic liquid crystals. *Mol. Phys.* **77**, 521 (1992).

[126] Jadzyn J and Czechowski G. The shear viscosity minimum of freely flowing nematic liquid crystals. *J. Phys.: Condens. Matter* **13**, L261 (2001).

[127] Iannacchione GS, Crawford GP, Zumer S, Doane JW, and Finotello D. Randomly constrained orientational order in porous glass. *Phys. Rev. Lett.* **71**, 2595 (1993).

[128] Dadmun MD and Muthukumar M. The nematic to isotropic transition of a liquid crystal in porous media. *J. Chem. Phys.* **98**, 4850 (1993).

[129] Iannacchione GS and Finotello D. Specific heat dependence on orientational order at cylindrically confined liquid crystal phase transitions. *Phys. Rev. E* **50**, 4780 (1994).

[130] Crandall KA, Rosenblatt C, and Aliev FM. Ellipsometry at the nematic-isotropic phase transition in a confined geometry. *Phys. Rev. E* **53**, 636 (1996).

[131] Cloutier SG *et al.* Molecular self-organization in cylindrical nanocavities. *Phys. Rev. E* **73**, 051703 (2006).

[132] Sheng P. Phase transition in surface-aligned nematic films. *Phys. Rev. Lett.* **37**, 1059 (1976).

[133] Steuer H, Hess S, and Schoen M. Phase behavior of liquid crystals confined by smooth walls. *Phys. Rev. E* **69**, 031708 (2004).

[134] Cheung DL and Schmid F. Isotropic-nematic transition in liquid crystals confined between rough walls. *Chem. Phys. Lett.* **418**, 392 (2006).

[135] Bellini T, Radzihovsky L, Toner J, and Clark NA. Universality and scaling in the disordering of a smectic liquid crystal. *Science* **294**, 1074 (2001).

[136] Qian S, Iannacchione GS, and Finotello D. Critical behavior of a smectic-A to nematic phase transition imbedded in a random network of voids. *Phys. Rev. E* **57**, 4305 (1998).

[137] Ondris-Crawford RJ, Crawford GP, Doane JW, and Zumer S. Surface molecular anchoring in microconfined liquid crystals near the nematic-smectic-A transition. *Phys. Rev. E* **48**, 1998 (1993).

[138] Kutnjak Z, Kralj S, Lahajnar G, and Zumer S. Calorimetric study of octyl-cyanobiphenyl liquid crystal confined to a controlled-pore glass. *Phys. Rev. E* **68**, 021705 (2003).

[139] Heidenreich S, Ilg P, and Hess S. Boundary conditions for fluids with internal orientational degrees of freedom: Apparent velocity slip associated with the molecular alignment. *Phys. Rev. E* **75**, 066302 (2007).

[140] Stark H. Saturn-ring defects around microspheres suspended in nematic liquid crystals: An analogy between confined geometries and magnetic fields. *Phys. Rev. E* **66**, 032701 (2002).

[141] Greulich S. *Rheologie von n-Alkanen in Nanoporen*. Diploma thesis, Saarland University, Saarbrücken, Germany (2007).

[142] Potter RW and Clynne MA. The solubility of the noble gases He, Ne, Ar, Kr, and Xe in water up to the critical point. *J. Solution Chem.* **7**, 837 (1978).

[143] de Gennes PG. On fluid/wall slippage. *Langmuir* **18**, 3413 (2002).

[144] Baranenko VI *et al.* Solubility of helium in water. *At. Energ.* **66**, 407 (1989).

[145] Markham AE and Kobe KA. The solubility of gases in liquids. *Chem. Rev.* **28**, 519 (1941).

[146] Clever HL, Battino R, Saylor JH, and Gross PM. The solubility of helium, neon, argon and krypton in some hydrocarbon solvents. *J. Phys. Chem.* **61**, 1078 (1957).

[147] Hesse PJ, Battino R, Scharlin P, and Wilhelm E. Solubility of gases in liquids. *J. Chem. Eng. Data* **41**, 195 (1996).

[148] C. Schäfer (private communication) .

[149] Hirama Y, Takahashi T, Hino M, and Sato T. Studies of water adsorbed in porous Vycor glass. *J. Colloid Interface Sci.* **184**, 349 (1996).

[150] Cruz-Chu ER, Aksimentiev A, and Schulten K. Water-silica force field for simulating nanodevices. *J. Phys. Chem. B* **110**, 21497 (2006).

[151] Mezger M *et al.* High-resolution in situ x-ray study of the hydrophobic gap at the water-octadecyl-trichlorosilane interface. *Proc. Nat. Acad. Sci. U.S.A.* **103**, 18401 (2006).

[152] Weast RC. *CRC Handbook of Chemistry and Physics* (CRC Press, Florida, 1981), 62nd edition.

[153] Wohlfarth C and Wohlfarth B. *Surface tension of pure liquids and binary liquid mixtures*, volume 16 of *Landolt-Börnstein - Group IV Physical Chemistry: Numerical data and functional relationships in science and technology* (Springer, Berlin, 1997).

[154] Pestov S. *Physical and Thermodynamic Properties of Liquid Crystalline Substances*, volume 5A of *Landolt-Börnstein - Group VIII Advanced Materials and Technologies* (Springer, Berlin, 2003).

[155] Langevin D. Light scattering from the free surface near a second order nematic to smectic A phase transition. *J. Phys. France* **37**, 901 (1976).

[156] Small DM, ed. *The physical chemistry of lipids: From alkanes to phospholipids*. Handbook of Lipid Research (Plenum Press, New York, 1986).

[157] Wilhoit RC, Marsh KN, Hong X, Gadalla N, and Frenkel M. *Densities of aliphatic hydrocarbons: Alkanes*, volume 8B of *Landolt-Börnstein - Group IV Physical Chemistry: Numerical data and functional relationships in science and technology* (Springer, Berlin, 1996).

[158] Wohlfarth C and Wohlfarth B. *Viscosity of pure organic liquids and binary liquid mixtures: Pure organic liquids*, volume 18B of *Landolt-Börnstein - Group IV Physical Chemistry: Numerical data and functional relationships in science and technology* (Springer, Berlin, 2001).

[159] Huber P and Knorr K. Adsorption-desorption isotherms and x-ray diffraction of Ar condensed into a porous glass matrix. *Phys. Rev. B* **60**, 12657 (1999).

[160] Page JH *et al.* Adsorption and desorption of a wetting fluid in Vycor studied by acoustic and optical techniques. *Phys. Rev. E* **52**, 2763 (1995).

[161] Saam WF and Cole MW. Excitations and thermodynamics for liquid-helium films. *Phys. Rev. B* **11**, 1086 (1975).

[162] Wallacher D, Huber P, and Knorr K. Adsorption isotherms and infrared spectroscopy study of nitrogen condensed in porous glasses. *J. Low Temp. Phys.* **113**, 19 (1998).

[163] Schäfer C. *Kalorimetrische Untersuchungen an mesoporösen Systemen*. Diploma thesis, Saarland University, Saarbrücken, Germany (2007).

[164] Mörz S. *Thermodynamische Untersuchungen an Porenkondensaten*. Diploma thesis, Saarland University, Saarbrücken, Germany (2009).

[165] Gruener S. *Fluss-Experimente an mesoporösen Membranen*. Diploma thesis, Saarland University, Saarbrücken, Germany (2006).

[166] Knudsen M. Die Gesetze der Molekularströmung und der inneren Reibungsströmung der Gase durch Röhren. *Ann. Phys.* **28**, 75 (1909).

[167] Smoluchowsky M. Zur kinetischen Theorie der Transpiration und Diffusion verdünnter Gase. *Ann. Phys.* **33**, 1595 (1910).

[168] Karniadakis G, Beskok A, and Aluru N. *Microflows and Nanoflows: Fundamentals and Simulation* (Springer, New York, 2005).

[169] Makri PM, Romanos G, Steriotis T, Kanellopoulos NK, and Mitropoulos AC. Diffusion in a fractal system. *J. Colloid Interface Sci.* **206**, 605 (1998).

[170] Hofmann T, Wallacher D, Huber P, and Knorr K. Triple point behavior of Ar and N_2 in mesopores. *J. Low Temp. Phys.* **140**, 91 (2005).

[171] Wallacher D and Knorr K. Heat capacity study on the melting of Ar in nanopores. *J. Phys. IV France* **10**, Pr7 (2000).

[172] Wallacher D and Knorr K. Melting and freezing of Ar in nanopores. *Phys. Rev. B* **63**, 104202 (2001).

Publications & Honors

Publications in Reviewed Journals

[1] Huber P, Gruener S, Schaefer C, Knorr K, and Kityk AV. Rheology of liquids in nanopores: A study on the capillary rise of water, n-hexadecane and n-tetracosane in mesoporous silica, *Eur. Phys. J. Special Topics* **141**, 101-105 (2007).

[2] Gruener S and Huber P. Knudsen diffusion in silicon nanochannels, *Phys. Rev. Lett.* **100**, 064502 (2008).

[3] Gruener S, Hofmann T, Wallacher D, Kityk AV, and Huber P. Capillary rise of water in hydrophilic nanopores, *Phys. Rev. E* **79**, 067301 (2009).

[4] Gruener S and Huber P. Spontaneous imbibition dynamics of an n-alkane in nanopores: Evidence of meniscus freezing and monolayer sticking, *Phys. Rev. Lett.* **103**, 174501 (2009).

Publications in Preparation

[5] Kusmin A, Gruener S, Henschel A, de Souza N, Allgaier J, Richter D, and Huber P. Polymer dynamics in nanochannels of porous silicon: A neutron spin echo study. (*submitted*)

[6] Gruener S and Huber P. Capillary rise and imbibition of liquids in nanoporous matrices: Rheological concepts and experiments. (*submitted*)

[7] Gruener S, Wallacher D, and Huber P. Influence of wettability and dissolved gases on the velocity boundary condition of Newtonian liquids in nanopores. (*manuscript*)

[8] Kusmin A, Allgaier J, Richter D, Gruener S, Henschel A, Huber P, and Holderer O. A neutron spin echo study of n-hexatriacontane and poly (ethylene oxide) confined in mesopores. (*manuscript*)

[9] Gruener S, Hermes HE, Egelhaaf SU, and Huber P. Capillary filling of mesoporous silica: Dynamics of imbibition front roughening. (*manuscript*)

Honors

- Diploma prize 2007 of 'Stiftung des Verbandes der Metall- und Elektroindustrie des Saarlandes' in recognition of the best diploma thesis at the physics department of Saarland University in the academic year 2006/2007 (2007).

- Post-doctoral fellowship of the German Academic Exchange Service for a two-year research visit in the group of Howard A. Stone at Princeton University, New Jersey, USA (2010).

Acknowledgement

It is impossible to acknowledge all of the contributions and support which I have received while working on the research behind this thesis and any attempt at full inclusion would certainly be lacking. Nonetheless, there are some contributions which are prominent in my mind.

Among the people I have to thank, my advisor, Patrick Huber stands out. Patrick has been a consistent source of intellectual wealth and his inquisitive nature is contagious. Furthermore, he has always been willing to lend an ear as well as provide suggestions during times of intense technical and physical challenge. However, it is his direction and kind words of encouragement especially during difficult times as well as his frank and sagacious advise that I am most appreciative of.

Special thanks is dedicated to my colleagues René Berwanger, Anke Henschel, Christof Schäfer, Volker Schön, and Matthias Wolff for the entertaining last five years. Thank you for making my time memorable and deeply rewarding.

I'd also like to thank all members of the working groups Huber and Wagner that I have had the pleasure of working with but also the administrative and technical team: Elke Huschens, Evelyn Treib, Karin Kretsch, and Rolf Kiefer as well as the in-house workshop under the direction of Michael Schmidt and the electronics technicians Jürgen Hoppe and Stefan Loew.

Furthermore, I am also appreciative of the contributions of Mikko Alava, Tommy Hofmann, Andriy Kityk, Klaus Knorr, and Dirk Wallacher. They all have significantly contributed to the intellectual content of this thesis. In the area of neutron radiography the co-workers Helen Hermes and Stefan Egelhaaf from Düsseldorf have been extremely supportive and helpful.

I acknowledge all of the various support provided me by my family. To my brothers and sister and to my mother for their love and friendship, to Markus and Vanadis for their love and encouragement, to Judy and Maggi for being there, thank you all. Finally, of all the contributions, my partner's stands out as the most meaningful. Her understanding and steadfast love has always been a source of profound tranquility during this sometimes hectic journey, . . .

. . . thank you Vivi!